Energy-efficient building design in Southeast Europe

Vladimir Jovanović

Energy-efficient building design in Southeast Europe

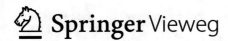

Vladimir Jovanović
Heidelberg, Germany

ISBN 978-3-658-24164-3   ISBN 978-3-658-24165-0   (eBook)
https://doi.org/10.1007/978-3-658-24165-0

Library of Congress Control Number: 2018964572

Springer Vieweg
© Springer Fachmedien Wiesbaden GmbH, part of Springer Nature 2019
Parts of this work were previously published in printed form as a dissertation by the author in the University
Library of the TU Vienna.

This Springer Vieweg imprint is published by the registered company Springer Fachmedien Wiesbaden GmbH
part of Springer Nature.
The registered company address is: Abraham-Lincoln-Str. 46, 65189 Wiesbaden, Germany

Preface

The aim of this book is to outline the principles of energy-efficient building design for Southeast Europe, as exemplified by housing in Serbia. This will involve considering the lessons learnt from traditional architecture, refurbishing issues associated with existing building stock, and the use of modeling tools to assess the performance of new designs. The intention is to develop a framework for reaching a positive energy balance in future buildings in Southeast Europe.

The audience to whom this book is aimed includes architects, urban planners, engineers, and other professionals in the building industry who are dealing with energy-efficient buildings. Other persons who would benefit from it are students of architecture, building sciences, and sustainability; investors interested in building developments in Southeast Europe, in particular in Serbia, along with scholars and individuals dealing with built environment issues.

The principles for energy-efficient design in Southeast Europe will be delivered in the form of conceptual design patterns that can be used as a starting point for designing a building. These will be backed up with scientific methodologies and data, which will be relatively easy to apply to any other region, especially to the rest of Southeast Europe. The results will be delivered in the form of graphs, diagrams, and appropriate tables outlining the findings from the use of the employed simulation tools.

The case studies were done during the period 2011–2015 for the example of Serbia. Serbia's location in the heart of Southeast Europe, covering geographical regions that spread far beyond the country's borders, gives it a unique status as a valuable case study for the whole region.

The building sector is a major user of energy, often in very inefficient ways. Therefore, integrating energy efficiency in the early-design phase of a building, be it a new structure or retrofitting, is essential in order to reduce the serious environmental impact of buildings, the support of long-term sustainable development, and ultimately to create a better future for us and coming generations.

This book is a result of research on energy-efficient architectural design, undertaken during the author's PhD studies at the Vienna University of Technology, and has been developed from the resulting doctoral dissertation.

I wish to thank my family, Ana, Miodrag, Slađana, and Milena, for their support during the course of my studies, and in the production of this book. I thank the Alfred Toepfer Stiftung F.V.S., Hamburg, for their support. I also thank my colleagues and friends, and especially my mentors for their contributions and guidance. In particular, thanks to Professors Karin Stieldorf, Manfred Berthold, and Klaus Kreč of the Vienna University of Technology and Professor Milica Jovanović Popović of the University of Belgrade.

Heidelberg, Germany   Vladimir Jovanović

Contents

List of Abbreviations and Acronyms

A/V	Area to volume ratio
AMSL	Height above mean sea level
AT	Austria
BIPV	Building-integrated photovoltaics
BPIE	Building Performance Institute Europe
C	Celsius
c	Thermal capacity
CHP	Combined heat and power
CO2	Carbon dioxide
COP	Coefficient of performance for heat pumps
d	Thickness
DE	Germany
DH	District heating
DHW	Domestic hot water
EC	European Commission
EPBD	Energy Performance of Buildings Directive
EPS	Extruded polystyrene insulation
ERV	Energy recovery ventilation
Etics	External thermal insulation composite systems
EU	European Union
EUR	Euro, the official currency of the euro zone
EW	Euro-Waebed simulation software
GHG	Greenhouse gas
HDD	Heating degree days
i	Investigation/Case study
i1R1	Investigation/case study 1, refurbishment scenario 1
K	Kelvin
kg	Kilogram
kWh	Kilowatt hour

kWh/m^{2}a	Kilowatt hour per square meter annually
LE	Low-energy standard
NEEAP	National Energy Efficiency Action Plan
NV	Night ventilation
NZEB	Nearly zero-energy building
PH	Passive house, Passivhaus
PHI	Passivhaus Institut
R	Refurbishment
RES	Renewable energy sources
RS	Serbia
RSD	Serbian Dinar, the official currency of the Republic of Serbia
S	South
SEE	Southeast Europe
t	Ton
UN	United Nations
U-value	Overall heat transfer coefficient
η	Energy conversion efficiency
λ	Lambda value, thermal conductivity

About the Author

Vladimir Jovanović
Born: 09.11.1986 in Banja Luka, Bosnia and Herzegovina

Since 2016, Consultant, Business Unit Architecture, io-consultants, Heidelberg, Germany
Licensed architect, Chamber of Architects Baden-Wuerttemberg, Germany
LEED AP (Accredited professional) BD+C, USGBC
Project Management Professional, PMP® PMI
2015, Energy Design Studio, Alpha Immobilien Consulting, Bamberg, Germany
2015, Doctor of technical Sciences (Dr. techn.), TU Vienna, Austria
2013, Herder-Scholar, Alfred Toepfer Stiftung F.V.S. Hamburg u. Vienna University
2011, Master of Architecture (MSc. Arch), University Niš, Serbia
2010, Bachelor of Architecture (BSc), University Niš, Serbia

Introduction

1.1   The Aims and Outline of This Book

High energy consumption has been emphasized as a major issue in the energy sector in Southeast Europe by the European Union (EU). Energy consumption, be it for heating, cooling, or other day-to-day activities, is also one of the major household expenses. This affects the budget of the home's occupants, who would need a sufficient in order to provide comfortable living conditions and well-being. At the same time, this high level of energy consumption, in a very direct way, influences the quantity of greenhouse gases being emitted to the atmosphere.

The EU directive EPBD Recast[1] has defined the year 2020 as the deadline for all newly erected buildings to be "nearly zero-energy buildings" (nZEB[2]). Although the average solar irradiation levels in Southeast Europe are higher than the European average (specifically, in Serbia it is 40% greater), architects and planners design mostly conventional buildings. This has led to the building stock being responsible for up to 50% of total energy consumption in Serbia, due to the inefficient use and choice of energy, and the inadequate energy efficiency of buildings. This is directly affected by the lack of available integrated energy-efficient designs. Considering these facts and issues, it is obvious that there is a need to improve local design procedures in order to deliver more energy-efficient buildings so as to meet the nZEB targets.

The aim of this book is provide a framework for integrating energy efficiency principles into the conventional building design process. The book focuses on thermal properties of buildings and especially on climate-appropriate designs for building envelopes, within the

---

[1]https://ec.europa.eu/energy/en/topics/energy-efficiency/buildings

[2]https://ec.europa.eu/energy/en/content/national-nearly-zero-energy-buildings-nzebs-definitions

© Springer Fachmedien Wiesbaden GmbH, part of Springer Nature 2019
V. Jovanović, *Energy-efficient building design in Southeast Europe*,
https://doi.org/10.1007/978-3-658-24165-0_1

context of Southeast Europe, although drawing upon examples of house design from different geographical regions of Serbia.

These challenges will be dealt with by this book by considering the following:

- To review traditional housing, which differs in the different parts of Serbia
- To examine the potential for enhancing energy efficiency in the current Serbian building stock by proposing refurbishment measures that are tailored to specific building types
- To propose design patterns for future housing that are appropriate for the different geographical regions of Serbia

The presented research was designed following a linear analytic structure. The primary methods employed were (1) case study research and (2) combined strategies (Groat and Wang 2002). The research tactic was predominantly experimental-based, making use of building simulation tools.

The target audience of this book are primarily architects, urban planners, engineers, and students, who are engaged in or are interested in the design and construction of more energy-efficient buildings. The intention is to provide a source that will help fill some of the gaps in knowledge that hinder the better application of energy efficiency solutions. It is hoped that this book will help those engaged in the design and construction of residential buildings in Southeast Europe (or anywhere for that matter), develop the skills required to interpret the characteristics of each region, and to be able to create an appropriate design that meets the local needs. Architects can and should improve their efforts in integrating sustainability into the project development and delivery process. As will be discussed, there is a great potential for Serbian architects to embrace such concepts. Brković and Milošević (2012) summarize this need well where they say "The evaluation of sustainable architecture in Serbia reveals that there is much more room for improvement. The question of broad and holistic understanding of all three aspects of sustainable design must be dealt with."

The resulting designs will be proposed via a simplified set of options, which have a flexibility and ease of application that allow an architectural style with a greater degree of energy efficiency. These choices, furthermore, are not dependent upon the actual style of the architect. Rather, they may be considered a series of guidelines that aim to help to design energy-efficient residences appropriate for the local environment in a cost-efficient manner. The guidelines are also developed with an open concept in mind, where the building itself may be upgraded as new developments arise, for example, the incorporation of improved photovoltaics, leading to the hoped-for achievement of zero-energy and energy-plus levels of efficiency.

Within the Southeast European region, Serbia has a unique central position, as it is composed of a number of different geographical areas that extent far from country's borders. This makes it good test case, since it can be said to be representative of Southeast Europe as a whole. Building energy behavior throughout the country will be examined within the context of indicators that reflect the environmental, economic, and social characteristics of such housing. These indicators will in turn be related to energy consumption, carbon footprint, the potential use of renewable energy sources (e.g., as done here, solar energy), investment return, and the actual level of environmental comfort that may be attained.

The book is in five chapters. This chapter, in addition to outlining what is hoped to be achieved, presented some general background regarding the push toward greater energy efficiency in buildings within a Southeast European context, especially with regard to Serbia. The following three chapters constitute the main body of the book. Chapter 2 deals with a review of traditional housing design in different parts of Serbia. The reason for this is that such dwellings were constructed within specific environmental and social contexts. This means not only the local climate but also the available resources needed to be considered. Today's architects would do well to heed the lessons of such builders when proposing their present-day designs. Chapter 3 is concerned with refurbishing current building stock, both historical and more modern. Retrofitting existing buildings offers a great deal of potential to reduce energy usage, although this must be understood within a social and economic context, which will also be touched upon. Chapter 4 presents a framework for incorporating more energy-efficient designs into contemporary housing. The framework is presented as an open concept toolbox, where specific characteristics of a building and its potential contribution to energy saving are outlined, taking the different environments of Serbia into account. The patterns and the guidelines presented in the conceptual form will help to design buildings more efficiently by suggesting appropriate and cost-effective actions throughout the design process. A series of investigations or case studies are presented from five locations around Serbia, namely, Subotica, Belgrade, Užice, Kopaonik, and Niš. The work considered renewable energy sources as well, namely, solar energy units, the potential of which in Serbia is great. The conclusions can then potentially be applied throughout Southeast Europe, given Serbia's position in the region.

The book concludes with a summary of these findings, as well as proposing avenues for future studies. The hope is that it will help to create a general mind-set from which architects and other professionals may employ the information contained within for improved and more energy-efficient and environmental-friendly designs.

1.2   The Energy-Efficient Economy

One of the major concerns of contemporary society, from economic, social, environmental, and scientific perspectives, is climate change. One way this manifests itself is in the recorded rise in average global temperatures as a result of increasing concentrations of GHG, primarily a result of anthropogenic activities (Kromp-Kolb and Formayer 2005). As a result of these concerns, the European Commission has committed itself to transforming the European Union to a more efficient, low-carbon emission economy (EC 2014a). The planned targets of this commitment are shown in Table 1.1, where by 2050 the aim is to reduce GHG emissions by between 80 and 95% with respect to 1990 levels. This is intended to be accompanied by both a general reduction in energy consumption and the

Table 1.1 EU Climate Policy targets projected for 2020, 2030, and 2050

Defined deadline	Reduction in GHG emissions (%)	Reduction in energy consumption (%)	Total energy made up by RES (%)
2020	20	20	20
2030	40	27	27
2050	80–95	Not defined	Not defined

Source: EC (2014b, c)

increased proportion of energy needs being made up by renewable energy sources (RES). For example, by 2020, energy consumption is planned to be reduced by 20%, while the share of total energy being provided by RES is aimed to be 20%, with a reduction in GHGs of 20%. By 2030, these values are expected to increase to a reduction in energy use of 27%, the proportion of energy usage by RES of 27%, with a reduction in GHGs of 40%.

One of the major sources of GHG emissions is associated with buildings, whose share of the EU's total energy consumption is some 40% (EU Legislation 2010). For example, within the recent Energy Performance of Buildings Directive (https://www.emissions-euets.com/energy-performance-of-buildings-directive-epbd) (EPBD, Directive 2010/31/EU), the concept is introduced of the nearly Zero-Energy Building (nZEB), which refers to a building that has a very high level of energy efficiency, and covers its zero or low energy demands to a significant extent by energy from renewable sources.

A crucial role in the development of energy efficiency in buildings is the contribution of architects, who have a major impact on building design. There are numerous challenges that must be confronted when designing nZEBs, especially when considering the targets set by the EU discussed above. In particular, there are gaps in the knowledge of the relevant professionals, especially in terms of keeping up with the ever changing standards and requirements that must be considered when engaging in these activities (BPIE 2011). The difficulties that are encountered in setting out to meet these goals are even a problem for leading economies such as Germany, where, as commented upon by the Building Performance Institute Europe (BPIE), "To date, one percent of all new buildings in Germany are built according to the passive house standard, therefore it can be assumed that at the EU level the percentage is smaller than one percent. The factor by which that should increase is therefore bigger than 100. A question that arises is whether the number of architects and installers that are able to deal with new technologies and standards will (have to) increase to satisfy the demand or not" (BPIE 2011).

1.3   Energy-Efficient Buildings in the EU

Energy efficiency in buildings is a critical issue for all current and future EU members, which has seen a number of studies funded by the EU. Within the European context, a huge potential for a more energy-efficient environment already exists with regard to the refurbishment of buildings. This is an area where stakeholders may play a part, ranging from the home owners themselves, designers, and craftsmen and builders (Gabriel and Ladener 2009). Refurbishment also needs to be applied to legally protected historical buildings, which despite the obvious difficulties involved, may achieve much greater energy efficiency via modern methods and technologies.

The assessment of the energy efficiency of buildings in Europe is strongly influenced by the country of concern, although most European countries have adopted the German passive house definition (Passivhaus Institut, Darmstadt, Germany). A passive house involves a space heating demand of 15 kWh/m^{2}a, with a heating load that does not exceed 10 W/m^2. In the event that active cooling is needed, the cooling energy demand again is limited to 15 kWh/m^{2}a. The total energy usage considering DHW (domestic hot water), heating, cooling, as well as auxiliary and household electricity for passive housing is limited to 120 kWh/m^{2}a. Furthermore, they are required to have controlled mechanical ventilation with heat recovery so as to meet the predefined air tightness level. Air tightness must in turn be verified by using an"n50" pressure test, where the required value must be less than 0.6 air changes per hour (PHI 2013).

However, there is a great variation between countries (even within federal states "Länder," as in the case of Germany) in terms of national building codes and nongovernmental initiatives (see Table 1.2 for some examples). When considering a passive house, the key factors are high-level insulation and window glazing, thermal-bridge-free construction, the maintaining of optimal environmental temperatures using the minimum space heating, air

Table 1.2 Some definitions of high-performance energy-efficient buildings in German-speaking countries (BPIE 2011)

Country	Definition of buildings	Requirements for energy efficiency
Austria	Klima:aktiv house	Heating demands: 70% of minimum requirements, correspond to 25–45 kWh/m^{2}a Heating demands for low-energy social buildings: maximum 60 kWh/m^{2}a (final energy consumption)
	Klima:aktiv passive house	Heating demands: 20% of minimum requirements, correspond to 15 kWh/m^{2}a Primary energy: 65 kWh/m^{2}a
Germany	Low-energy building (KfW40)	Energy demands: 40% of minimum requirements (according to the building code)
	Passive house	Heating demands < 15 kWh/m^{2}a Primary energy < 120 kWh/m^{2}a Heating load < 10 W/m^2
Switzerland	"A" labelled building	Energy demands: 50% of minimum requirements
	Minergie	Delivered energy for dwellings: 38 kWh/m^{2}a For industry: 20 kWh/m^{2}a; for offices: 30 kWh/m^{2}a
	Minergie-P	Delivered energy for dwellings: 30 kWh/m^{2}a (dwellings) For industry: 15 kWh/m^{2}a; for offices: 25 kWh/m^{2}a

tightness, ventilation with heat recovery to maintain good indoor air quality, proper orientation with respect to the sun, overheating prevention, solar hot water systems, insulated
ventilation ducts and DHW pipes, and energy-efficient household appliances and lighting
(BRE 2011).

1.4   The Republic of Serbia: An Introduction

1.4.1   Geography and Climate

The Republic of Serbia is a landlocked country located in the heart of the Southeast Europe
(see Fig. 1.1). It has an area of 88,407 km², of which 24% is made up of the autonomous
province of Vojvodina, with Kosovo and Metohija[3] making up another 12.3%. It has a
population (excluding Kosovo-Metohija) based on the most recent census in 2011 of
7,120,666 (Serbian Government 2014).

   As mentioned above, Serbia has a varied geography that extends far throughout the rest
of the region into its eight neighboring countries: Hungary in the north, Romania and
Bulgaria to the northeast and east, respectively, Macedonia and Albania to the south,
Montenegro in the southwest, and Bosnia and Herzegovina and Croatia in the west.
North part of Serbia, Vojvodina, is a mainly flat area located on the Pannonian Plain.
The country's central part and the hilly Šumadija region are located to the south of the Sava
and Danube Rivers, with the Danube being the country's longest river. Moving southward,
the hill country gradually becomes more mountainous, with peaks reaching altitudes of
over 2000 meters above the sea level. Forested areas make up a total of around 27.3% of the
country.

   Serbia has a moderately continental climate with some local variations. Most of the
country lies within a temperate climate zone, although the southwest borders continental
and subtropical zones. There is some seasonal variability, with January being the coldest
month with mean monthly temperatures (depending on the region) of −6 to 0°C, July the
hottest with mean monthly temperatures ranging from 11 to 22°C, and autumn usually
being warmer than spring. In terms of rainfall, most of the country experiences continental
patterns with an annual average over the country of 896 mm. Generally, most rain falls in
the warmer half of the year, except for the southwest, which experiences higher rainfall in
autumn. Snow is common from November to March, while in mountain areas above
1000 m, there may be snow all year round (TOS 2014).

   Because of this varied geography extending through the region, Serbia may be used as a
prime case study for the development of more energy-efficient building designs. The
specifics of a given region will need to be considered in the design process, which in

---

[3]Based on United Nations Security Council Resolution 1244, Kosovo and Metohija have been under
the temporary civil and military administration of the United Nations since 10 June 1999.

Fig. 1.1 Map of Central and Eastern Europe. Serbia (marked bottom-left on the map) is positioned in the center of the Balkan Peninsula where it is made up of geographical regions that extend throughout Southeast Europe (Own illustration, based on original map from United Nations 2008, Map No. 3877, Rev. 7)

turn implies that such designs could be applicable to areas with similar geographical characteristics.

1.4.2   Energy and Housing

The total primary energy consumption[4] of Serbia in 2011 was 16.2 Mtoe (megatons oil equivalent). This was provided mainly by coal (54%), oil (23%), gas (8.6%), biomass

---

[4]http://ec.europa.eu/eurostat/statistics-explained/index.php/Glossary:Primary_energy_consumption

(6.4%), and hydroelectricity (4.6%). The final energy consumption[5] for 2011 was estimated to be 9.25 Mtoe, divided between coal with 12.2%, oil with 34.6%, gas with 8%, electricity with 25.2%, district heat with 9% (i.e., a distributed heating system delivered via district networks, from various sources), and RES with 11%. In terms of the final energy consumption by the different sectors, residential, services and agriculture made up 48.9% of usage, industry 29.4%, and transport 21.7% (Austrian Energy Agency 2013).

An environmental aspect of Serbia that has the potential to contribute significantly to its renewable energy resources is the relatively high annual number of sunshine hours. For most regions, this is between 2000 and 2500 h (around 40% greater than the European average, Stamenić 2010). Such levels offer great opportunities for cost-savings in energy usage, especially if this contribution is incorporated as part of a wider, integrated energy-efficient design process. For example, the BPIE (Building Performance Institute Europe[6]) believes savings as high as 80% may be made in terms of a building's operational costs if appropriate integrated designs are followed, of which solar energy would form part. This would also involve little (if any) extra cost over the lifespan of the building. However, Serbia is very much behind other EU member states in terms of the exploitation of solar energy. Hence, one of the aims of this book is to outline how this direction can contribute to enhanced building energy efficiency and reduced GHG emissions.

Residential buildings in Serbia number around 2,246,320, leading to a total of 3,188,000 dwellings (National Typology of Residential Buildings in Serbia). Because of the inefficient thermal properties of these structures, these buildings lead to a very high level of energy consumption (Jovanović Popović and Ignjatović 2013a, b), which has been raised as a major concern. The residential sector in fact is one that stands to provide the most gains in terms of improved energy efficiency. For example, the First National Energy Efficiency Action Plan (NEEAP) comments that the residential sector in Serbia has the greatest potential for cost-saving through greater energy efficiency. This sector is responsible for around 38% of the country's total energy consumption (Todorović 2010), where energy usage for heating and domestic hot water is around 220 kWh/m^2 per year (Implementation Program of the Energy Development Strategy of the Republic of Serbia). In terms of electricity usage, households consume around 56% the national total, 65% of which goes toward heating, which forms the largest proportion of energy consumption in Serbia (Belgrade Chamber of Commerce 2010). Figure 1.2 shows the proportion of dwellings in Serbia by the employed heat source, where households mostly use local stoves, district heating system, and electric heaters.

---

[5]http://ec.europa.eu/eurostat/statistics-explained/index.php/Glossary:Final_energy_consumption
[6]http://bpie.eu/about/who-we-are/

Fig. 1.2 The percentage of
dwellings in Serbia classified by
the used heat source. Source:
Energy Agency Serbia

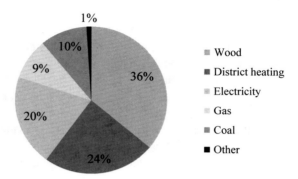

Heat sources for households

1.5   Energy Policy in Serbia

Serbia has already greatly improved upon its policies with regard to energy efficiency,
taking into consideration the importance of the associated political and legal frameworks.
Such frameworks need to take into account advances in renewable energy, planning and
construction, as well the continued refinement, or even replacement, of existing
regulations. All of these issues add additional complexity to any energy policy. This has
lead Serbia to adopt a number of strategic documents and laws in relation to policies related
to energy use and efficiency. Considering first international frameworks, Serbia became a
member of the Energy Community[7] in 2006. This was a first step toward implementing
European Union regulations (Pucar 2012), accepting the target of obtaining 27% of its total
energy needs from renewable sources by 2020 (EC 2014a, b). At the same time, this will
see the harmonization of its energy policies with those of the EU. In addition, the Energy
Community is helping with the preparation of guidelines to deal with the most important
policy issues. It is also supporting the development of action plans and strategies, such as
the National Energy Efficiency Action Plans and National Renewable Energy Action Plans
(see below). Serbia has also ratified the Kyoto Protocol as part of its contributions to efforts
to protect the environment. One aspect of the Kyoto Protocol is the provision for the
transfer of clean technologies from more developed countries to less developed ones.

The main features of Serbian energy policy deal with energy efficiency, competitive
energy markets, increased use of renewable energy sources, and the protection of the
environment. These aspects are embedded into the so-called Serbian Energy Law and the
Energy Development Strategy (Law, Energy 2014). As implied above, national action plans

---

[7]https://www.energy-community.org/. The Energy Community is an international organization that
sets out to bring together the European Union and its neighbors to create an integrated pan-European
energy market.

are being developed within the context of broader international frameworks, such as above mentioned the Energy Community. Furthermore, and referring specifically to the building sector, energy efficiency is dealt with within several regulatory frameworks, namely, the Law on Construction and Planning, the Rulebook on Energy Efficiency in Buildings (Ministry of Energy, Development and Environmental Protection of the Republic of Serbia 2011), the Rulebook on Energy Certification, and the Law on Efficient Energy Use.

Considering first the new Law on Energy (dating from 2014), this legislation is responsible for determining the long-term goals of the energy sector's development, building upon the previous 2012 Law on Energy. Included in this law is the full liberalization of the Serbian electricity and gas markets, and the harmonization of the energy sector of Serbia with EU regulations through the implementation of the EU Third Energy Package (Law, Energy 2014). All of these laws and regulations are meant to allow the integration of the Serbian market with the wider EU. One major aim of the law was the establishment of conditions whereby improved energy efficiency and the development and exploitation of renewable energy sources were encouraged. This aim in turn resulted in the Decree on Incentive Measures for the Production of Electricity from Renewable Energy Sources and Combined Heat and Power Production (Law, Energy 2014), which involved the introduction of a feed-in tariff regime in 2009 and updated in 2013.

With regard to the Energy Development Strategy, until 2015, this policy focused on the more sensible use of better-quality fuels and increased efficiency with regard to energy production, distribution, and end-user consumption. However, several barriers to achieving these aims were pointed out by the supplementary Strategy Implementation Programme 2007–2012. The first involves the unrealistic parity in energy prices and their variability (i.e., there are different tariffs within the pricing of electricity, depending on the time of consumption, day/night, etc.). The second was how energy in Serbia had been assumed to be social category for decades, a consequence of its communist past, which has led to cheap electricity (compared to the rest of Europe), resulting in high consumption and low efficiency. The current strategy, the Energy Development Strategy to 2025 with projections to 2030, was entered into force in 2016 (Ministry of Mining and Energy of Republic of Serbia 2016). The issues covered by the strategy include developing Serbia's energy production capacity, and improving the transmission, transport, and distribution system, all while ensuring the country's energy security and at the same time protecting the environment.

The first of the abovementioned National Energy Efficiency Action Plans of the Republic of Serbia (NEEAP) was adopted in 2010 and dealt with the period from 2010 to 2012 (Todorović 2010). This plan included various targets, namely, the reduction of final energy consumption during the period 2010–2012 by 1.5%, or 0.1254 Mtoe (in fact, preliminary results have shown savings of 1.2% savings), and further savings of 9% in final energy consumption by 2018. The Second NEEAP was submitted in 2013 with the goal for the period 2013–2015 of 3.5% savings in comparison to 2008 (0.2952 Mtoe). The Third NEEAP was published in 2018, with the goal for the period 2016–2018: 4.6% savings in comparison to 2008, or 0.3824 Mtoe (Energy Community 2017).

The National Renewable Energy Action Plan (NREAP) was responsible for setting targets for the increased use of renewable energy sources by 2020, as well as proposing means by which these targets may be met and how to enhance investment in such ventures. In addition to the aim of having 27% of the gross final energy consumption generated from renewable energy sources by 2020, the distribution of the use of renewables is planned to be 30% in heating and cooling, 36.6% in electricity consumption, and 10% in green transport (Ministry of Energy, Development and Environmental Protection of the Republic of Serbia 2013).

Finally, the Spatial Plan of the Republic of Serbia 2010–2020, which led to the Law on Spatial Plan of the Republic of Serbia 2010–2020, consists of frameworks for the strategies being developed for spatial planning, renewable energy use and development, energy efficiency, and ecology. The main problems highlighted by this plan include the lack of reliable data on the potential of renewable energy resources, the lack of programs and projects that would attract investment, issues about some of the renewable energy resources themselves (e.g., how some of them are dependent upon climatic conditions), unequal distribution network, relatively low current electricity prices which lessens the motivation for investment into energy-saving measures, and the often unclear ownership relationships. While the challenges facing Serbia in terms of developing a more energy-efficient and environmental-friendly building stock are rather serious, there has been considerable effort expended in creating a policy and political framework that would encourage such an undertaking.

This chapter explained the aims and outline of this book, the importance of energy efficiency in buildings, and an introduction of the geography, climate, and energy policies of Serbia. The following chapter will review the traditional architecture of Serbia. Such structures and styles are the product of both human (history, migration, etc.) and environmental (climate, geography, available resources) factors. It is therefore essential for designers to understand the background, to both learn from successes, and to build upon what works, and what can be improved.

References

Austrian Energy Agency. (2013). Total energy consumption by sector – Serbia 2010. https://www.enercee.net/index.php?id=306 (accessed 10.04.2013).

Belgrade Chamber of Commerce. (2010). Sectoral Collaboration Project with Regard to Financing Energy Efficiency in Buildings within the Frame of EU Regulations and Legal Arrangements, Country report: Serbia. EUbuild Energy Efficiency, IPA Project, country report Serbia. Retrieved from http://www.eubuild.com/wp-content/uploads/2011/06/9CountryReport-SERBIA1.pdf (accessed 10.02.2013).

BPIE. (2011). Principles For nearly Zero-energy Buildings: Paving the way for effective implementation of policy requirements. Brussels: Building Performance institute Europe.

BRE. (2011). Passivhaus primer: Designer's guide A guide for the design team and local authorities. Retrieved from http://www.passivhaus.org.uk/page.jsp?id=108 (accessed 15.02.2012).

Brković, M., & Milošević, P. (2012). Architects' perspective on sustainability in Serbia: Establishing key topics. Spatium, (28), 60-66.

EC. (2006). Directive 2006/32/EC of the European Parliament and of the Council of 5 April 2006 on energy end-use efficiency and energy services and repealing Council Directive 93/76/EEC. Official Journal of the European Union, 27, 136–148.

EC. (2014a). EU action on climate, retrieved from http://ec.europa.eu/clima/policies/brief/eu/, (accessed 10.11.2014).

EC. (2014b). The 2020 climate and energy package, retrieved from http://ec.europa.eu/clima/policies/package/index_en.htm (accessed 10.11.2014).

EC. (2014c). 2030 framework for climate and energy policies, retrieved from http://ec.europa.eu/clima/policies/2030/index_en.htm, (accessed 10.11.2014).

Energy Community. (2013). Report on the implementation of the Energy Performance of Buildings Directive in the Republic Of Serbia, Directive 2010/31/EU on the Energy Performance of Buildings (EPBD).

Energy Community. (2017). retrieved from https://www.energy-community.org/dam/jcr:62046fe5-6473-47b8-96e6-2ed9354b7ae1/EECG032017_Ministry%20of%20Mining%20and%20Energy.pdf , (accessed 15.07.2018).

Energy Agency Serbia. (2011). Regulation of electricity prices, retrieved from http://www.aers.rs/FILES/Prezentacije/2012-12-3%20EPS%20RegulCena%20dec%202012%20LJM%20c.pdf, (accessed 10.12.2012).

EU Legislation. (2010). Summaries of EU legislation, Energy performance of buildings, retrieved from http://europa.eu/legislation_summaries/internal_market/single_market_for_goods/construction/en0021_en.htm, (accessed 15.11.2014).

Gabriel, I., & Ladener, H. (2009). Vom Altbau zum Niedrigenergie+ Passivhaus. Gebäudesanierung. Neue Energiestandards. Planung und Baupraxis.

Groat, L., & Wang, D. (2002). Architectural research methods. New York.

Jovanović Popović, M., & Ignjatović, D. (2012). Atlas of Family Housing in Serbia. Faculty of Architecture, University of Belgrade and GIZ–German Association for International Cooperation, Belgrade.

Jovanović Popović, M., & Ignjatović, D. (2013a). National Typology of residential buildings in Serbia. Faculty of Architecture, University of Belgrade and GIZ–German Association for International Cooperation, Belgrade. Retrieved from http://www.arh.bg.ac.rs/wp-content/uploads/201415_docs/SAS_EEZA_publikacije/National_Typology_of_residential_buildings_in_Serbia.pdf, (accessed 10.12.2014).

Jovanović Popović, M., & Ignjatović, D. (2013b). Atlas of Multi Family Housing in Serbia. Faculty of Architecture, University of Belgrade and GIZ–German Association for International Cooperation, Belgrade.

Kromp-Kolb, H., & Formayer, H. (2005). Schwarzbuch Klimawandel: wie viel Zeit bleibt uns noch?. Ecowin-Verlag.

Law, Efficient Energy Use. (2013). Official Gazette of the Republic of Serbia No 25/13. Belgrade.

Law, Construction and Planning. (2011). Official Gazette of the Republic of Serbia No 72/09, 81/09 – correction, 64/10-US и 24/11. Belgrade

Law, Energy. (2012). Official Gazette of the Republic of Serbia No. 57/11, 80/11-amendment, 93/12 and 124/12). Belgrade.

Law, Energy. (2014). Official Gazette of the Republic of Serbia No. 145/2014. Belgrade.

Ministry of Energy of the Republic of Serbia. (2005). Energy Sector Development Strategy of Republic of Serbia by 2015. Official Gazette of the Republic of Serbia, No. 35/05. Belgrade.

Ministry of Energy of the Republic of Serbia. (2007). Energy Sector Development Strategy Implementation Programme 2007–2012. Official Gazette of the Republic of Serbia, No. 17/07 and 73/07. Belgrade.

Ministry of Energy, Development and Environmental Protection of the Republic of Serbia. (2013). National Renewable Energy Action Plan. Belgrade.

Ministry of Energy, Development and Environmental Protection of the Republic of Serbia. (2011). Rulebook on energy efficiency of Buildings. Official Gazette of the Republic of Serbia no. 61/2011. Belgrade.

Ministry of Energy, Development and Environmental Protection of the Republic of Serbia. (2009). Decree on incentive measures for the Production of Electricity from Renewable Energy Sources and Combined Heat and Power Production. Official Gazette of the RS, No. 99/09. Belgrade.

Ministry of Mining and Energy of Republic of Serbia (2016), retrieved from http://www.mre.gov.rs/latinica/dokumenta-efikasnost-izvori.php (accessed 15.07.2018).

PHI. (2013). Passivhaus, retrieved from http://www.passivhaus-institut.de/, (accessed 20.01.2013).

Pucar, M. (2012). Building and Regulations in the Field of Energy Efficiency and RES in Serbia, Regional Countries and EU. Network Conference 2012, Kecskemét, Hungary.

Serbian Government. (2014). Population, retrieved from http://www.srbija.gov.rs/, (accessed 15.11.2014).

Stamenić, Lj. (2010). Alternativna energija Srbije, Ekonomist Magazin br. 518 spec. dodatak, Politika a.d.

Todorović, M. (2010). First NEEAP/BS national energy efficiency action plan/building sector 2009-2018. u: Study Report and NEEAP-BS for the Republic of Serbia Ministry of Mining and Energy. Washington: IRG, June.

TOS. (2014). National Tourism Organisation of Serbia, retrieved from http://www.serbia.travel/about-serbia/facts/, (accessed 10.03.2014).

UN. (2008). Central and Eastern Europe. Department of Field Support, Cartographic Section. Map No. 3877 Rev. 7.

UN. (2014). A world of Information. Retrieved from https://data.un.org/, (accessed 10.06.2014).

Traditional Homes

2

The first question to ask before embarking on efforts to develop improved, more energy-efficient designs in the future is to consider how and why house designs varied between regions in the past. To answer this question, this chapter will review the diverse range of traditional dwellings found in the different parts of Serbia. The following discussion of folk architecture in Serbia makes use of the historical-interpretative method. The results of the summary in turn helped in selecting the locations for the detailed evaluations discussed later in this book.

The chapter begins by describing the characteristics of Balkan settlements, along with descriptions of typical houses in towns and rural areas. This will include explaining the relevant peculiarities of each region's traditional housing and presenting some representative examples of each type of architecture.

2.1 Balkan Settlements

The territory of what is present-day Serbia has been settled since ancient times, with prehistoric settlements frequently positioned in the same vicinity as modern settlements. The Balkans have seen many types of settlements established as a result of the complex history of the region, with influences including those from the Mediterranean, Western Europe, and the Ottoman Empire (Cvijić 1922).

While there have been studies on the archaeology and monumental architecture of Medieval Serbia, there is very little information on rural and popular structures. Rural settlements were generally built a distance from military roads owing to the frequent wars and general insecurity among the inhabitants. The houses were simple affairs made from relatively weak materials such as mud, wattle, wood, and branches. This led them to having fairly short life spans, which explains the lack of information about them in the literature.

© Springer Fachmedien Wiesbaden GmbH, part of Springer Nature 2019   15
V. Jovanović, *Energy-efficient building design in Southeast Europe*,
https://doi.org/10.1007/978-3-658-24165-0_2

The first studies into traditional Serbian housing, which may be considered the basis for emerging research into folk architecture, was by Cvijić (1922) who examined architecture from the seventeenth and eighteenth centuries. It is seen that from the seventeenth century onward, no one unique style which may be proclaimed as being typically Serbian existed, owing to the many factors that influenced housing across the country. These include, firstly, the noticeable differences between houses in towns and those in rural areas. Such a difference is not surprising when one considers the fact that these two grouping arose under different social conditions, where in the towns the houses were built to satisfy the needs of those who lived and worked in the town, whereas in rural areas, dwellings developed according to the materials that were locally available, the nature of the terrain and environment, and the livelihood of the occupants.

2.2   Old Towns

Serbian cities trace their origins mainly from Roman and Byzantine times, although some may be even older. There was also a very strong influence from the Ottoman Empire. For instance, in Central and West Serbia, people tended to live in smaller rural settlements rather than in cities. These settlements eventually evolved into townships or, to use the Serbian word, *varošica*. They subsequently developed into linear layout settlements, which was modeled after those in Vojvodina, and planned following Austro-Hungarian regulations. Furthermore, old towns in Kosovo and Metohija generally retained their original Balkan structure. In terms of the distribution of the population, during the Ottoman siege of Serbia, Serbs mostly lived in villages, while Turks, Greeks, and Jews lived in towns (Cvijic 1922).

A typical small township in the Balkans was made up of many short, narrow, and curved streets, along which were numerous small buildings and stores. The residences were usually built outside of the center of the town, within fenced courtyards. Larger townships had central squares and markets. However, old towns in Serbia usually also had straight and wide streets, which led to the market which was the focus of trade and commerce for the town, houses were usually small, with only a ground floor, surrounded by large back-yards which were used for gardens.

2.2.1   Traditional Balkan Houses in Townships

From the fifteenth century onward, towns in Serbia developed a new form of architecture, which was an extension of the commonly used Byzantine style as well as including an oriental element. The term used to define this new style is *Balkan profane architecture* (Kojić 1949). Kojić stated that "Balkan profane architecture emerged from the long period of combining and mixing impacts of Byzantine civilization and Turkish characteristics." A representative example of this architectural style was the so-called traditional Balkan house, a common form found throughout all the towns on the Balkan Peninsula during the eighteenth and nineteenth centuries. Cvijić described this style in a similar way, where he noted "The house survived many mixed and combined impacts, which were afterwards hard to define and identify clearly."

Considering similarities between the old Byzantine style of housing and the Balkan house, the main common feature was in the organization of space, where the Balkan house most likely retained the original features of earlier Byzantine houses. These features included a compact floor plan, two to three stories, and a large and high-ceiling central hall with many windows and openings, around which the other rooms were arranged. The Turkish influence was gradual and is generally considered to be secondary features. For example, the interior of the house gained many new decorative elements, with façades being built with many windows, oriels, porches, and verandas. The modifications themselves, it must be kept in mind, were modeled on the styles of oriental buildings and incorporated the needs and culture of that society's conservative family life. However, differences also arose between the old Byzantine and Balkan houses in terms of the choice of materials used. For the Byzantine culture, stone and wood were used for housing, whereas in Medieval Serbia, while stone was used for monuments, it was used much less in houses which were mainly made of wood and mud. Regarding the traditional Balkan town house (see Fig. 2.1), these had a foundation of stone, a ground floor made of stone or in a half-timbered system, and an upper floor build as a half-timbered construction filled in with wattle and daub or with adobe bricks (Kojić 1949).

A traditional Balkan town house appeared to have a very high second floor when compared to the ground floor. Its walls would appear to extend outside of the ground floor's edge, a result of the upper floor being separated from the ground floor by a visible wooden slab. This led to the ground floor having a large width/height ratio and relatively small windows (see the example in Fig. 2.1). Horizontal shaping was the dominant architectural form; for example, the upper floor façade was made up of a horizontal row

Fig. 2.1 Traditional Balkan house in the town of Pirot, Serbia (Museum of Ponišavlje in Pirot, with permission of Čedomir Vasić)

of windows, while the supporting columns at the façade's edges were painted in dark colors and visually emphasized. The street façade often had one or two oriels (protruding bays), while the back façade always had an open porch. The façades were treated with plaster, but did not have any other additional decoration or artistic motives. The houses were painted white to enhance the shade effect of the deep eaves, while the eaves themselves were deep overhanging structures, up to 2 m deep in southern parts of the country, with additional strut stays for reinforcing the structure. Roofs had a pitched form with low slopes and were covered by tiles, while the chimneys were slim with decorative caps rising high above the house. These houses were surrounded by a green courtyard closed in by a high fence and a large entrance gate (Kojić 1949).

Traditional Balkan houses were built as detached or terraced dwellings. Rich home-owners would build detached dwellings with a large courtyard and a garden. If it was constructed on a slope, then the building would be positioned in order to make best use of the vista. On the other hand, terraced houses were built in dense town centers (Deroko 1974). In addition, two distinct types of traditional Balkan house in Serbia were to come about, namely, the symmetrical and asymmetrical types (Kojić 1949).

2.2.2 Symmetrical Type of Houses in Towns

The symmetrical type of town house, which was often a two-story building, was the style mainly built by the Turks. These buildings displayed a symmetry in both their floor plans and façades. The floor plan was characterized by a main central hall surrounded in a regular form by other rooms. The street façade often had two mirrored bay windows, while the back façade had one central porch facing the courtyard (Fig. 2.2).

Fig. 2.2 The symmetrical type of town house in Serbia (Stambolijski House in Niš). The construction of this house started about 1875. Stambolijski House belongs to Balkan profane architectural style (see above), with more than 10 separate, spacious rooms (Adopted from niscafe.com)

2.2.3 Asymmetrical Type of Houses in Towns

The asymmetrical type of town house (Fig. 2.3) was characterized by the specific positioning of a porch toward a garden. The porch was not in the middle, but rather it was displaced toward the edge of an open longitudinal veranda and elevated one step above the ground. An entrance from the veranda led directly into the largest room. The house's rooms were not of an equal size, and there was no central hall, but there was an open fireplace (in Serbian: *odžaklija*), often located in the middle of the largest room, but not always. This room also served as the kitchen during the winter months. Houses of the asymmetrical type were mostly one-story buildings, although there were also two-story examples. A similarity between these asymmetrical houses and the rural "Morava house" (which will be described later, see Fig. 2.8) can be observed. The reason for this similarly is probably because the asymmetrical houses were mainly built by Serbs who settled into towns and who designed their new homes after the Morava houses of their villages.

In summary, the traditional Balkan house in a town was representative of the Balkan profane style of architecture. However, the Balkan house could not really be described as being representative of regional architecture in Serbia for a number of reasons. First, a town's architecture often followed a new architectural style that had been previously seen in another developed town, regardless of the nature of its location. This was noticeable in southern and northern Serbia, where either Turkish or Western European influences dominated, respectively. Secondly, architecture in towns was not wholly dependent upon local resources. For example, a citizen who dealt with craft and trade would find it much easier to obtain building materials from a nearby township. Finally, climatic conditions did not significantly affect the type of construction undertaken with regard to either the shape or the general appearance of the traditional Balkan house. Balkan profane architecture may therefore be found in towns and cities in other countries, such as Bosnia and Herzegovina, Montenegro, Macedonia, Albania, Bulgaria, Turkey, and Greece.

Fig. 2.3 The asymmetrical type of town house in Zaječar, East Serbia (with permission of Čedomir Vasić)

2.3   Rural Settlements

The rural areas in Serbia had a more diverse development than the townships. For example, as mentioned above, during the Ottoman period, Serbs mainly lived in villages. These villages differed in their character as a result of how they developed. Cvijić (1922) defined a classification scheme for villages from the seventeenth to eighteenth centuries, referring to dense villages, middle dense villages, and less dense villages (in Serbian: *sela zbijenog tipa, sela srednje zbijenog tipa i sela razbijenog tipa*).

The less dense type of villages was found in the western and central parts of Serbia. Their main feature was that the houses were positioned relatively distant from each other, where the space between rural households was used as agricultural and farming land. The households or family communities were made up of a house, auxiliary buildings, and a large yard for the growing of vegetables and fruit. Such communities therefore presented a small rural economic system. Based on the character of these villages, they are further classified into Stari Vlah villages, Šumadija villages, Mačva and Jasenica villages, and Ibar villages, after the regions where they be found.

On the other hand, dense villages, which were mostly found in Kosovo-Metohija and in the Morava Valley, were some of the densest villages in the Balkans. These villages had many curved streets, linked courtyards, and houses that were built close to each other. Dense villages grew outward as a kind of agglomeration. If there was no more space for expansion, for example, due to surrounding hills, rivers, and lakes, a new village emerged, few kilometers from the native/original settlement. These settlements were positioned in valleys close to rivers, but rarely in mountains. Again, these villages were classified further to Timok villages, Čitluk villages, Turkish villages, and mixed types.

Settlements in rural areas adapted to their environment, influenced by factors such as the terrain, climate, the availability of local materials, the knowledge of the local builders, and the typical housing styles in the area. This led, for instance, to mountain villages being very different to those located in valleys. As a result, a regional rural architecture emerged.

2.3.1   From Primitive Housing to Folk Architecture

Rural architecture in Serbia began with primitive housing such as huts. These dwellings had a circular basis and a conical shape. The round basis allowed for easy construction, with the covering make up of branches, leaves, hemp, and grass. It consisted of one round room with a fireplace in the center. There was only one small door, no windows or chimneys, no roof space (meaning no separate roof construction, e.g., no attic), and no floor covering, except for earth. Cvijić describes a number of types of these huts: *sibara, busača, kulača,* and *dubirog* or *savrda* (the last type had an opening in its roof for smoke to be released).

In the seventeenth and eighteenth centuries, a number of pit houses (dugout houses) were built in the Srem area, Vojvodina. Similarly, the Morava Valley saw one-room houses without a roof space, covered with hemp or wood bark, being built. In the beginning of the nineteenth century, mobile log cabins were a feature of Mačva and Sićevo Gorge. These log cabins were supported by stones and large logs, but had no foundations. They could therefore be disassembled for transport to another location, although the roof had to be transported in one piece.

The first phase of the development of log cabins in Serbia began in the nineteenth century. The initial predominant style of cabin had a square floor plan with one room, a fireplace in middle, and a roof covered with hemp or wood shingles. The walls were about 2 m high and were constructed from carpentry trimmed logs placed on a low stone foundation. The first and the last horizontal belt in the wall's height were from very wide logs, approximately 10 cm wide and 25 cm high. The doors were always toward one edge of the wall, not in the middle.

Eventually, the wattle and daub hut evolved into the wattle and daub house. A major advancement was the introduction of foundations. With foundations, walls received support, although struts were required to ensure the stability of the walls. The roof in turn was held together by logs and walls. Such a structure thus gained the main features of a real house. However, these primitive types of houses are not necessarily representative of Serbian folk architecture, but rather they reflect primitive housing in general.

The development of traditional houses in Serbia started from the eighteenth century. Before this time, the frequent wars that ravaged the region left many areas in Serbia abandoned. Therefore, the development of rural houses in Šumadija and the Morava Valley over the last three centuries represented the rapid development of housing from ancient styles until more modern times. This saw houses transforming from being simply primitive wood and mud shelters without walls, windows, or chimneys to fully developed houses with foundations, built walls, glass windows, built fireplaces and chimneys, plaster finishing, and decorative details.

Kojić (1949) describes the development of traditional homes in Serbia (except in Vojvodina) in four phases. First, in the beginning of the eighteenth century, the land was less settled, leading to large-scale migrations in the second half of the eighteenth century. The housing, while primitive, was in groups, hence forming the origin of present-day villages. Second, from 1804 onward (i.e., the time of the First Serbian Uprising, or as it known in Serbian, *Prvi srpski ustanak*), houses suddenly changed from being of a primitive form and began to evolve in terms of their shapes and means of construction. Due to the plentiful supply of wood from the huge forests, fully wooden houses emerged. Third, from 1820 onward (after the Second Serbian Uprising), regional architecture began to emerged. Log cabins dominated in the Drina Valley and Šumadija, while half-timbered buildings were predominant in Kosovo-Metohija, the Morava Valley, and Ponišavlje. Fourth, in the second half of the nineteenth century, different construction styles merged, leading to half-timbered construction to become dominant (largely due to the now lack of hardwood logs). These houses had three or four rooms, with some variation in their form and shape. This process of the evolution of traditional homes suggests that every generation made their own steps toward the development of their styles housing.

Consequently, three types of traditional homes have been noted as being the most specific representatives of folk architecture in Serbia. These are log cabins, half-timbered houses ("Morava houses"), and rammed earth houses, examples and their geographical distribution shown in Fig. 2.4.

2.3.2 Log Cabins in Šumadija, the Drina Valley, and Stari Vlah

A log cabin, termed *Osaćanka*, could be considered to be the purest representative of an architecture that developed in Šumadija, the Drina Valley, and Stari Vlah during the first half of the nineteenth century. The name originated from the well-known builders, *Osaćani*, who came from the settlement of Osat in Bosnia and Herzegovina, which was the origin of the Dinaric log cabins. Two factors played an important role in the emergence of this type of log cabin: (1) the availability of massive logs from the forests that could be used as the building material and (2) the need to ensure good shelter from the snow and cold of the harsh winter conditions (Fig. 2.5). Older log cabins were built using round logs, while later ones were made from half-circled logs, that is, the logs are cut lengthwise, leaving a semicircular profile.

The basic plan of an "Osaćanka" house was rectangular. In the interior there was a square room with an open fireplace and a wide vent for releasing the smoke. The main room (in Serbian: *kuća*) had wooden walls, a rammed earth floor, and a fully open fireplace. The fireplace was asymmetrically positioned, being displaced toward one of the walls. The main room usually lacked windows, or, if small windows existed, they were initially covered with animal skins. Light in the evening was provided by the fireplace or candles. The opening that served as the entrance to the main room or cabin was short, but relatively wide. A cabin often had two doors on opposite sides of the building, which allowed good natural ventilation. The upper part of the entrance door often displayed a decorative wooden arc carved into a softwood. On the inside there were no ceiling finishes, with only rafters that joined at a ridge board. The roof was always covered with oak shingles and was high and steep, as can be seen in Figs. 2.5 and 2.6, preventing the buildup of too much snow during winter. This led to roofs being higher in the mountains and a little lower in Šumadija. There were no gable walls on the façades. A particular element of these cabins was a chimney which was a wide pipe that started from the ceiling, just above an open fireplace, and became thinner as it through the roof, before being topped with a decorative cap, called in Serbian a *kapić*. Later forms of these log cabins saw the division of the main room into a living room and a bedroom. The living room never had a ceiling, while it was common for the bedroom to have a wooden ceiling.

The building techniques employed in the "Osaćanka" log cabins at the time were at a rather high level. While the foundations were made from stone, the main building material was oak wood, which had to be treated carefully with hand tools. This leads to the joining often being done without any additional metal attachments. Not a single element in such a house was bought in a town, with the entire building being made from local materials. This

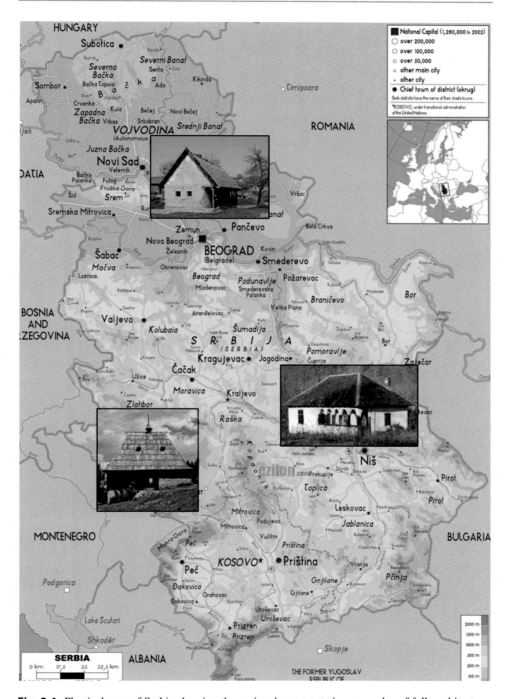

Fig. 2.4 Physical map of Serbia showing the regional representative examples of folk architecture. In the north there is the rammed earth housing of Vojvodina, to the west the log cabins, and to the east the Morava houses

Fig. 2.5 A log cabin, "Osaćanka," on a slope in the ethno village of Sirogojno (An ethno village is a common term in the Balkans for a traditional authentic village preserved for tourism, culture and education) (with permission of Open-Air Museum "Staro selo" Sirogojno)

Fig. 2.6 A log cabin. This example was the house of Knez Miloš Obrenović in the village of Gornja Crnuća and was where the decision about the "Second Serbian Uprising" was made, making it, in a way, a center of the state during the period 1815–1819. Left: Sketches of the floor plan, the section, the façade and the axonometric view. Right: A photograph from 2009 [Sketches adapted from Kojić (1949); with permission of IP Prosveta and of Čedomir Vasić]

includes features such as the door hinges, locks, and even the keys, which were made from a hard wood by local craftsmen.

The "Osaćanka" style of log cabin was a unique style in the Balkans, which until the end of the nineteenth century was the dominant type in Serbia, especially in the mountains and hills. It was typical for the forest areas of west and southwest Serbia, although a few examples may be found in the Morava Valley and east Serbia. They were also characteristic for Southeast Bosnia and Herzegovina and the hilly parts of Montenegro. However, while log cabins were a typical form of housing in forest areas across Europe, Asia, and North America, this type of housing has largely disappeared, since hard woods are more expensive than conventional building construction. This has resulted in these houses today only surviving in isolated mountainous and hilly areas.

2.3.3   Transforming the Log Cabin to Semi-wattle and Daub Houses

The chronological development of log cabins in Serbia according to Kojić (1949) may be described as consisting of four phases. The first was the one-room log cabin. These were closely related to forest areas and usually lacked a porch.

The second phase began with the addition of a storage room without windows, which later evolved into the heated room. This additional room was built using the wattle-daub technique and was heated by a "heater on pot" (in Serbian: *peć na lončiće*), although the main room was still the center of family life. The result was the dwelling becoming a semi-log cabin and a semi-wattle-daub house, where the main room, called the house, was always built from logs, and the additional room made from wattle and daub (Fig. 2.7)

The third phase saw the addition of a porch to the above described semi-log cabin and semi-wattle-daub house. The resulting three-room house replaced the shingled roof for a tiled one and the wooden chimney with one built of brick. This style of house had more windows, which were all glass covered. For family communities, there were additional auxiliary buildings, with a courtyard fenced by gates that were often decorated. More developed family communities had a well or *bunar* in Serbian in the yard. Less developed communities had to bring water in from a nearby spring.

The fourth and final phase involved moving toward two-story houses. This involved the emergence of the "house on a slope" (in Serbian: *kuća na ćelici*), in order to exploit the

Fig. 2.7   A house on a slope. This is the house of Vuk Karadžic in Tršić, built following a semi-log cabin and semi-wattle-daub technique (with permission of Čedomir Vasić)

sloped terrain of such areas. When the slope was steep, the lower level of the house was established in the ground so that it became a ground floor. Few two-story log cabins in Serbia were built, as wattle and daub (or adobe brick) rooms were less durable than pure log cabins, leading to semi-log cabins and semi-wattle-daub houses decaying relatively quickly when they were abandoned.

2.3.4 Half-Timbered "Morava House" in the Morava Valley

The half-timbered house was the style that was followed in the less harsh climates in Serbia. The most characteristic type was the "Morava house" (in Serbian: *Moravska kuća*). At the beginning of the nineteenth century, log cabins dominated in the Morava Valley, while by the end of the nineteenth century, Morava houses were adopted throughout all of southern and eastern Serbia, the Morava Valley, Jasenica, Lepenica, the Timok Valley, and Kosovo. The old Morava houses were constructed by builders from North Macedonia and South Serbia, such as "house builders from Crna trava" (in Serbian: *Crnotravci*). In the Sava and Danube Valleys, the Morava house retained its original form, but the building techniques differed, making use of the rammed earth house methods followed in the Pannonian Basin (Cvijić 1922).

Morava houses in Serbia were rather large and were strongly associated with dense and middle dense villages, in contrast to the log cabins of the more mountainous areas of West Serbia. Morava houses therefore did not have a large courtyard or many auxiliary buildings, but rather the courtyard was rather compact, employing the more efficient use of free space. Their interiors were generally divided into three rooms, namely, a main living room, a second room which served as the bedroom, and a porch (in Serbian: *trem*). The two rooms were almost equal in size, as in the semi-cabin and semi-wattle-daub houses. The main room had a semi-open fireplace that leaned against the internal wall, allowing the bedroom to be also heated, with the smoke leaving through a chimney with a decorative cap.

The porch was a key element of this style of housing, being made up of a fenced open entrance hall in front of the house, with rammed earth floor and wooden pillars supporting the roof eaves. Porches were also used as spaces for drying food and storing tools and formed the link between the house and the courtyard or a field where villagers would have carried out most of their daily activities. Sometimes they were not in the front of the façade, but just to one side, with a two-level terrace with a height difference of ca. 40–80 cm toward one side, originally called a *Doksat*. Old types of porches had regular wooden pillars, while the latter, more advanced Morava houses, had decorative arcades made from laths and mud between the pillars on the porch as a decorative and nonstructural element (Fig. 2.8).

The structure of these houses was made up of a frame built from light wood, filled in with wattle and daub, or adobe bricks. Adobe bricks were commonly used, especially in the southern areas, because there was ample solar radiation to allow for the sun-drying of clay.

Fig. 2.8 Decorative arcades in the "Morava style" on the Archeo-Ethno Museum in Ravna. This building was constructed at the end of the nineteenth and beginning of the twentieth century. Note how the terrace has a step on the left side, which is an example of a Doksat. However, such regular two-story houses with a half-timbered construction were rare in Serbia (with permission of Čedomir Vasić)

Fig. 2.9 Morava houses built in a half-timbered construction. The roofs are gradual, reflecting how there was less snow in these areas, with deep eaves, meant to protect the façades from rain and sun. The house in the center shows decorative arcades (with permission of Čedomir Vasić)

In addition, the façades, which were made from clay plaster, were protected from rain and the sun by deep eaves. The walls were usually painted white, both inside and out, and the roofs, which were four sided, were covered with fired clay tiles. Since there was less snow in the areas where Morava houses were build, the roofs were more gradual than, for example, the log cabins described above (Figs. 2.9 and 2.10).

Fig. 2.10 Old Morava house in village Razbojna (with permission of Nevena Brkic)

2.3.5 Rammed Earth Houses in Vojvodina

Vojvodina differs from other regions in Serbia, historically and geographically, in that it is a flat plain which is very suitable for agriculture with few forests and a great deal of groundwater. European influences in this region were always stronger than those from the Byzantine, and later the Turkish, empires. During the Ottoman era (1526–1699), Vojvodina was a poor area with pit houses (dugouts) and few villages. The pit houses had only an outer roof, or if they were situated on a slope, they were built fully underground, probably a result of the severe winters in this region. Those pit houses in flatter terrains had a gable roof with gable openings for the entrance, daylight, and air circulation. Although this type of house was dominant in Vojvodina during the sixteenth and seventeenth centuries, there is evidence that small wattle and daub houses covered with straw with one or two rooms also existed here, often forming a type of semi-pit house. Gradually, the population started to build rammed earth houses with an organized yard, whose shape and size were closer to the later nineteenth-century houses of this area.

During the eighteenth century, the Austrian-Hungarian Empire instigated massive migration to Vojvodina, leading to the founding of new settlements and villages. These villages developed in a ribbon and orthogonal scheme, where on the main crossroads, there was a square around which were located the more important buildings. The resulting planned parcellation of the land required the houses to be positioned in a yard, which in turn affected the shape of the house itself, leading eventually to a new architectural style over the whole Pannonian Basin (Fig. 2.11).

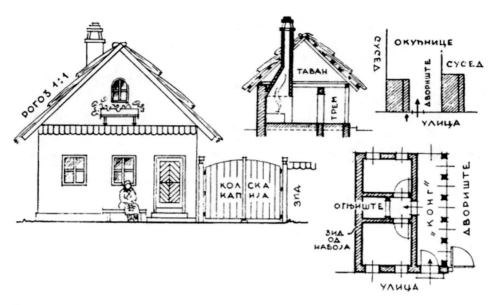

Fig. 2.11 Sketch of the frontal façade, section, floor plan, and courtyard plan of a rammed earth house in Vojvodina (from the left to the right) [Adapted from Kojić (1949), with permission of IP Prosveta]

Considering a village following a ribbon layout, each house is intended to have optimal access to both a street or road and a field. This leads to both the courtyards and the houses, which were usually one story, to be relatively long and thin. The positioning of the houses saw a frontal gable façade situated close to a street, which was painted every year, usually in bright vivid colors. One of the side-façades would be attached to a neighboring house, while another side-façade was orientated toward the courtyard. The houses had a central kitchen with a semi-open fireplace. Above the fireplace there was a chimney that covered most of the room. In front of the kitchen there was one or two rooms, and all rooms were linked to the courtyard by a longitudinal porch. People entered the house directly from the street via this porch. To the back of the house, there was a stall for cattle, a garage, and other buildings for economic activities. Food and grain storage structures were situated toward the other neighbor's border.

Since earth was the main material available in Vojvodina, these houses were usually rammed earth constructions. The half-dried mud for the walls was reinforced with cut straw and then rammed in a timber frame, layer by layer. Once it became solid, the frame was removed. The roofs were always gabled and were covered at the early stages with reeds from nearby rivers, but later clay tiles were used (Fig. 2.12).

Subsequently, at the end of the eighteenth and the beginning of the nineteenth century, again following administrative policy, the development of villages in a ribbon pattern formed the final layout of contemporary villages in this region. Considering again the

Fig. 2.12 Rammed earth houses in Vojvodina. Left: typical positioning of the house in a village that developed along a ribbon plan. Right: the building materials employed can be seen in this picture, namely, mud and reeds (image courtesy of Nenad Glišić, with permission of "Moja Vojvodina, moj dom")

housing, it is apparent from the above discussion that the emergence of the rammed earth style of housing in Vojvodina was quite different from the previous two styles of Serbian folk architecture in Serbia, namely, the log cabin and the Morava house.

2.4   Summary

The review of folk architecture presented in this chapter described the characteristic housing of several different parts of Serbia. Considering first town architecture, it was noted that this did not vary much, regardless of the local climatic conditions. By contrast, rural architecture did show considerable differences between regions, a product of the populations there taking into careful consideration the local conditions, namely, topography, climate, and what resources and materials were available. In turn, these considerations dictated to a large extent the building techniques that were employed in these areas. Specific examples include the use of steep-sloped roofs in the hills of West Serbia, while more gradual roofs with deep sun-protecting eaves were adopted in the valleys of Central Serbia.

Furthermore, the three styles of folk architecture (see Fig. 2.4) in Serbia are unique: log cabins, rammed earth houses, and half-timbered "Morava houses." This summary of the styles and characteristics of traditional homes will be one of the key criteria when selecting the locations for the case studies presented later in this book. It also provides background information on the basis upon which climatically appropriate architectural styles in different regions of the country have been and will be developed.

In the next chapter, we will examine how the exiting building stock of Serbia, considering both historical and more modern buildings, can be retrofitted. Use will be made of simulation tools for a series of case studies, where not only architectural designs will be considered but also socioeconomic factors, choice of heating source, as well as the presentation of a design pattern toolbox.

References

Cvijić, J. (1922). Balkansko poluostrvo i južnoslovenske zemlje [Balkan Peninsula and the South Slavic Lands]. Beograd: Državna štamparija Srba, Hrvata i Slovenaca.

Deroko, A. (1974). Narodna arhitektura: Folklorna arhitektura u Jugoslaviji. Naučna knjiga.

Kojić, B. Đ. (1949). Stara gradska i seoska arhitektura u Srbiji. Prosveta.

Open air Museum Old Village Sirogojno. (2018). Retrieved from http://www.sirogojno.rs/en/about-museum (accessed on 15.07.2018).

Moja vojvodina, moj dom. (2018). Retrieved from https://www.facebook.com/pg/Moja-Vojvodina-moj-dom-107404402638681/about/?ref=page_internal (accessed on 15.07.2018).

Refurbishing Existing Building Stock

3

3.1 Introduction

The previous chapters provided a review of the evolution of the different types of traditional homes in Serbia, within their historical and environmental context. We saw how previous authors developed various approaches to describe this architecture, and how there was continuous innovation as new ideas for constructing houses were developed. This in turn provided the groundwork for the research on this topic which forms the bulk of this book, and what will be presented in the coming chapters, starting with existing building stock.

Why should we be concerned with existing building stock? The reason is that the energy efficiency of current buildings will play a major role in reducing energy usage and, as a consequence, GHG emissions. As recent studies in the EU have pointed out, around 80% of the current European building stock will still be in existence in the year 2050 (Neuhoff et al. 2011). Therefore, any consideration dealing with reducing energy usage while enhancing the energy efficiency of the future building stock must take into account existing buildings, hence requiring the analysis of their energy use behavior.

It is therefore the aim of this chapter to describe the research design followed that lead to the proposed design patterns outlined at later stages of this book. An outline of the evaluation criteria employed for the diverse building types examined is first presented, followed by a description of the simulation software used. Then an overview of the existing building stock of Serbia, including a discussion of the sources of energy used, their efficiency, and their contribution to GHG emissions, will be given.

The main part of this chapter will be a series of case studies dealing with issues surrounding the refurbishment of existing buildings. These studies will cover the refurbishment of a historical building (a very specific case), then residential housing from the 1970s and 1980s. Next, some socioeconomic barriers to retrofitting will be presented, including an

© Springer Fachmedien Wiesbaden GmbH, part of Springer Nature 2019
V. Jovanović, *Energy-efficient building design in Southeast Europe*,
https://doi.org/10.1007/978-3-658-24165-0_3

evaluation of opportunities for sustainable high-impact refurbishment. Next, design patterns for house refurbishment will be outlined. In this part, the focus will be on how to improve upon the building's thermal envelope and how this can be modified for improved energy usage while maintaining a high level of comfort. Finally, the efficiency of different heating systems used in Serbia and their environmental impact will be discussed. As a reference point, the energy saving potential of replacing the heating system and the amount of carbon dioxide emissions from different heat sources for one case study will be presented.

3.2   Evaluation Criteria

The first step in developing energy-efficient designs for new houses or for refurbishing existing ones is to identify the evaluation criteria that should be used in assessing energy efficiency. For the assessments presented in this book, Table 3.1 outlines the criteria that were employed. These cover a range of aspects, from energy efficiency, environmental impact, economic and socioeconomic factors, thermal comfort, and structural damage prevention. While the listed indicators do not consider all possible assessment factors (suggesting there is still considerable potential for further research in this field), the ones used nonetheless allow a multidimensional perspective for energy-efficient design in Serbia.

Considering these indicators in some detail, and one can see that to assess energy efficiency, the heating demands of the buildings in question need to be considered. Another related issue involves the potential for photovoltaic electricity, which will be considered via a case study later in this book. With regard to environmental impact, the amount of GHG emissions from a particular heat source, as well as the efficiency of the source itself, are indictors that can also be related to those dealing with the building's energy use and the potential of RES. Considering economic issues, including energy and investment costs, and the potential for financial returns on investments need to be investigated, which are related to socioeconomic factors such as local energy prices, resident income, and the attractiveness of investment. Internal thermal comfort is evaluated through the overheating potential during the summer season, while for specific cases, the prevention of structural damage was addressed by thermal bridge analysis.

3.3   Simulation Software Tools

To undertake the case studies, several software packages were used to assess the buildings' energy use and efficiency behavior, as well as to obtain other required data. The packages used were EuroWaebed, GEBA, AnTherm, GEQ, and PVGIS.[1]

---

[1]For software description see references at the end of the chapter.

Table 3.1   The evaluation criteria employed in this book for the investigated buildings

Energy efficiency and RES	Environmental impact	Economic issues	Local socioeconomic factors	Thermal comfort	Structural damage prevention
Heating demands of the building	Efficiency of heating devices	Energy costs, investment costs	Income of residents and the attractiveness of investment	Overheating potential	Heat flow and diffusion
Photovoltaic electricity potential	GHG emissions	Investment return period	Income to investment relation	Internal air temperatures	Thermal bridge analysis

The EuroWaebed (EW) tool estimates monthly and annual heating demands for initial basic building models (thermal properties that would be measured in a real building) and a variety of refurbishment scenarios (see, e.g., Bointner et al. 2012). It is able to accommodate factors such as heat gains from equipment, heat losses, the presence of occupants, and air infiltration. Climate data for the EW simulations were gathered from the PVGIS Solar Radiation Database (see below), which in turn made use of climatologic data from the European Solar Radiation Atlas.

The GEBA simulation software is employed to predict the thermal behavior of a building in summer by assessing the suitability of particular types of construction by, for example, evaluating potential overheating. This leads to the different structural types being tested with respect to the resulting internal comfort, specifically the internal air temperatures in rooms that may experience overheating. The climate parameters and conditions required for the simulations were defined after consultations with the tool's developers (Krec 2013).

The AnTherm simulation software calculated the temperature distribution within buildings, in particular heat flows and vapor diffusion. It is considered a reliable tool in terms of meeting the required European standards for the precise evaluation of a building's thermal performance (AnTherm 2014). For the calculation of the linear diffusion of vapor within building elements, the GEQ tool was also used. GEQ is an Austrian energy certification software (in German, *Energieausweis Software*), in accordance with the Austrian norm ÖNORM B 8110-2: 2003-07-01 (GEQ 2014).

PVGIS is a web-based tool for estimating the monthly and annual potential photovoltaic (PV) electricity generation of free-standing and building-integrated PV modules, developed and implemented by the Joint Research Centre from the European Commission's science services. PVGIS considers the tilt and orientation of the PV units and includes a Google-map application as its geographic information system (GIS) component, enhancing it user-friendliness.

3.4   Existing Building Stock and Energy Sources

3.4.1   Housing Stock

As discussed in the introduction to this book and pointed out by the NEEAP, there is great potential for improving energy efficiency in the Serbian residential sector (Austrian Energy Agency 2013). For example, this consumption sees on average around 228 kWh/m²a used for heating and domestic hot water preparation (Belgrade Chamber of Commerce 2010). With this in mind, this chapter presents some appropriate measures for the refurbishment of existing buildings in Serbia. The data used was collected from the Statistical Office of the Republic of Serbia,[2] and from the book 'National Typology of Residential Buildings in Serbia' (Jovanović Popović and Ignjatović 2013a). The refurbishment of existing buildings will be considered by undertaking multiple case studies, where two main target groups of buildings are investigated: (1) historical buildings built before 1919 and (2) various types of residential buildings built during the 1970s and 1980s.

Considering the first group, protected historical buildings are a particular group and require a specific set of methods and considerations when undergoing refurbishment. This is because the external appearance of these buildings should remain essentially unchanged. Hence, some of the main challenges when retrofitting historical buildings involve limiting modifications to the façades and dealing with the building's physical structure when upgrading the building's internal envelope. If not considered carefully or neglected, these issues could led to serious structural damage to the building.

Regarding the second group, the housing stock in Serbia was significantly enlarged during the 1970s and 1980s, with around 1,150,000 housing units, or 42% of total housing stock, coming into being within those 20 years (Statistical Office of the Republic of Serbia 2013). Most of the existing residential buildings are characterized by the poor thermal properties of the building envelopes. It is also worth mentioning that around 70% of the Serbian population lives in houses (i.e., not apartment blocks) which have a similar concrete structure with brick block walls and cement plaster. This also involves the frequent (although not always) deficiency of proper thermal insulation.

3.4.2   Sources of Energy and GHG

As energy efficiency is the central theme of this book, some discussion is required in the actual sources of energy in Serbia. It also needs to be discussed within a broader context than simply what a household consumes. For example, the quantity of energy consumed for space heating in households is much greater than the net "heating demands." Heating demands represents useful energy that has been converted and transported from the initial

---

[2]http://www.stat.gov.rs/en-US/

Table 3.2 Energy flow within the classification scheme employing the concepts of primary, secondary, and final energy

Energy form	Technology	Examples
Primary		Coal, wood, hydro, dung, oil, etc.
	Conversion	Power plant, kiln, refinery, digester
Secondary		Refined oil, electricity, biogas
	Transport/transmission	Trucks, pipes, wires
Final		Diesel oil, charcoal, electricity, biogas
	Conversion	Motors, heaters, stoves
Useful		Heat, shaft power

Source: Hulscher (1991)

primary energy source. Primary energy itself is the energy that is available from the natural environment. This leads to considering secondary energy, which is the energy expended on the transport or transmission of the energy. The end result is the final energy, which is what is actually received by households. In some cases, the primary energy could be at the same time the secondary and even the final energy (e.g., wood gathered for cooking or heating). The flow of energy following this description is outlined in Table 3.2 (Hulscher 1991).

Breaking down further the steps from final to useful energy for a households' space heating is important, since each conversion step results in some energy loss. Considering the input and output of energy for heating purposes, the final energy which households actually receive is dissipated during the generation of the heat, then during the distribution of the energy through a heating system, and finally in its delivery, leaving the useful energy as only a fraction of the input energy. This may be illustrated, as outlined in the Serbian regulations on energy efficiency in buildings, by the product of the generation (ηk), distribution (ηc), and delivery (ηr) of energy to give the energy conversion efficiency, η. This is a critical aspect to consider, since energy losses arising in its transport from production to final usage may range from 2 to 35%, dependent upon the heating source. The data about energy conversion efficiency of heating systems for individual houses are presented in Table 3.3.

Considering the issue of CO_2 emissions by buildings as a result of their energy consumption (a point raised at the earliest parts of this book), this is largely dependent upon the actual heating source. There are, of course, many different heating methods, each having their own advantages and disadvantages with respect to operational costs, environmental impact, and other factors.

Table 3.4 compares the CO_2 footprint of different heat sources in Serbia and compares these values with those arising from Croatia and Austria in order to gain a wider perspective of how efficiency varies across the broader region.

Table 3.3 Energy conversion efficiency (η) of heating systems for individual houses in Serbia

Phase	Source	η	Additional information
Generation: furnaces, boilers	Solid fuels	0.65–0.68	Devices up to 50 kW, without regulation or manual
	Oil	0.81–0.83	Devices up to 50 kW, manual regulation
	Gas	0.80–0.88	Devices up to 100 kW
Distribution: piping	Local	0.95–0.98	Non-insulated and insulated pipes in the envelope
	District heating	0.88–0.92	Pre-insulated pipes from district heating system
Delivering regulation system		0.90–0.95	Without zoning of the building
		0.92–1.00	Zoning

Source: Serbian Rulebook on Energy Efficiency in Buildings

Table 3.4 CO_2 emissions from different heat sources from Serbia, Croatia, and Austria

Energy source	Serbia	Croatia	Austria
Natural gas	0.20 kg/kWh	0.20 kg/kWh	0.238 kg/kWh
Liquefied petroleum gas	0.215 kg/kWh	0.215 kg/kWh	–
Extra-light oil	0.265 kg/kWh	0.265 kg/kWh	–
Light oil	0.28 kg/kWh	0.28 kg/kWh	0.312 kg/kWh
Biomass (wood)	–	–	0.015 kg/kWh
District heating	0.33 kg/kWh	0.33 kg/kWh	0.039–0.074 kg/kWh
Hot water (geo./ conv.)	–	–	0.059–0.295 kg/kWh
Electricity	0.53 kg/kWh	0.53 kg/kWh	0.40 kg/kWh
Coal (domestic)	0.32 kg/kWh, brown coal	1.5 kg/kg, hard coal	–
Coal (imported)	0.40 kg/kWh, brown coal	1.88 kg/kg, hard coal	0.34 kg/kWh
Lignite (domestic)	0.33 kg/kWh	1.0 kg/kg	–
Waste heat	–	–	0.02 kg/kWh
Electricity (imported)	–	–	0.648 kg/kWh

Source: Serbian Rulebook on Energy Efficiency in Buildings; Croatian Rulebook on Energy Efficiency in Buildings; Austrian OIB Richtlinien

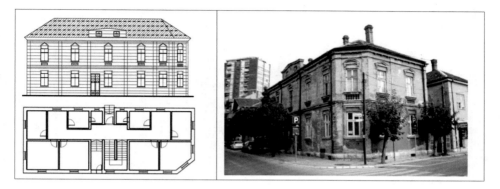

Fig. 3.1 The historical building built before 1919 used as the example for case study 1. This example, located in the center of the city of Niš, Serbia, was built in 1909. Left: the ground floor and the front façade. Right: a photograph of the building from 2013

3.5 Case Studies and Refurbishment Scenarios

3.5.1 Case Study 1: Historical Building Renovation

The example used in this case study, which is concerned with evaluating the measures proposed for refurbishing historical buildings built before 1919, is located in the center of the city of Niš, Southeast Serbia (note, this location experiences 2613 heating degree days[3] per year). The example selected is typical of such buildings, being built in 1909. Fig. 3.1 shows the example with the floor plan and façade view on the left-hand side and a photograph of the façade as of 2013 on the right-hand side.

The building has a total heated-floor area of 437.92 m^2, where both the ground and upper floors have an area of 218.96 m^2. The basement and attic are not heated. The external walls were built from clay bricks with plaster finishing, with the total façade area being 537.16 m^2. The ceiling height was 3.57 m, with a timber-framed construction. The roof, which was covered with clay tiles, was also timber framed with four gradual slopes. The total window area was 58.48 m^2, where 26.56 m^2 each faced the southeast and southwest directions, each. The windows were double glazed (80% being glass and 20% wooden frames). The entrance doors had an area of 4.2 m^2. Most importantly to this case, the building was not insulated and had a low-energy performance.

---

[3]A heating degree day is a metric for quantifying the amount of heating buildings in a given location will need over a certain period. One heating degree day represents conditions outside a building were equivalent to being below a defined threshold comfort temperature inside the building by one degree for one day.

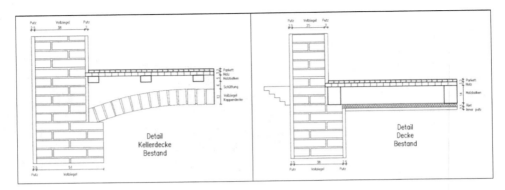

Fig. 3.2 The structure of the joints (left) between the external wall and basement ceiling (detail 1) and (right) between the external wall and the ceiling (detail 2)

The case study involved assessing the heat energy saving potential of the building, how to improve thermal comfort, in particular the internal temperatures during summer, and how to prevent structural damage during and after the refurbishment. In addition, a thermal bridge analysis was performed for two structural (joint) elements, namely, the joint between the external wall and the basement (termed detail 1) and that between the external wall and the ceiling (detail 2) (see Fig. 3.2). Other assessments were made for heat flow and vapor diffusion in the building elements and the need to ensure that the proposed refurbishment suggestions are suitable for such a historical building in terms of not leading to any structural damage.

For the purpose of this case study (and those following), an experimental simulation approach was employed. Four simulation tools were employed: (1) EW for calculating the annual heating demands; (2) GEBA to evaluate possible overheating in summer; (3) AnTherm was used to analyze the heat flow and surface temperatures for retrofitted building elements; and (4) GEQ calculated the linear vapor diffusion within the building's elements. To undertake the simulations, a detailed description of the building and its various components was required, including their thermal properties. The thermal properties of the original and the retrofitted building elements are summarized in Table 3.5.

In order to preserve the appearance of the façade, internal insulation was applied, leading to limited refurbishing measures being carried out. This involved the addition of several insulating elements, namely, 5 cm of chip board "WS-EPS" was used for the external walls, 5 cm of mineral wool and 8 cm of expanded perlite for the basement ceiling, and 10 cm of EPS insulation above the ceiling in the roof. In addition, airtightness was improved by replacing the old doors and the old windows with double-glazed (4-12-4 mm) windows. How effective these refurbishment efforts were is shown in

Table 3.6 in terms of the reduced heating demands and the internal air temperatures during summer. The thermal comfort was evaluated by comparing the change in the temperature of a room oriented toward the southwest for July, where it would have a tendency to overheat. The resulting temperatures prior to and after refurbishing are presented in Fig. 3.3.

Table 3.5 Overall heat-transfer coefficients of building elements, where the U-value is the heat-transfer coefficient

Building element	Basement ceiling	External walls	Roof ceiling	Windows	Frames	Doors
U-value [W/m²k] original	1.28	1.61	1.02	2.91 (6-16-6 mm)	3.5	3.5
U-value [W/m²k] renovated	0.29	0.46	0.30	1.10 (4-12-4 mm)	1.8	1.8

Table 3.6 Heating demands and improved thermal comfort after the proposed refurbishment

Scenario	Refurbishment measures	Heating demands	Summer conditions (internal air temperatures)
i1-R1[a]	External walls—5 cm of "WS-EPS" chip board Windows and doors—(4-12-4 mm) Roof ceiling—10 cm EPS. Basement ceiling—5 cm wool and 8 cm perlite Airtightness improved	Initially: $q = 195$ kWh/m²	Initially: 34.7 °C
		Refurbished: $q = 63$ kWh/m²	Refurbished: 32.0 °C

[a]Note: Here and in the rest of this chapter, the terminology to indicate a particular case study and scenario is expressed as iX-RY, where iX refers to case study/investigation X and RY refers to retrofitting scenario Y. Hence, as this is the first case study and first scenario, it is denoted by i1-R1

The main challenges involved when retrofitting historical buildings are associated with the fitting of the internal insulation and dealing with issues such as thermal bridges, moisture and mould growth. Therefore, the two structural elements described in Fig. 3.2 were evaluated using the AnTherm simulations. Figure 3.4 shows the initial condition of these elements, with the expected results from the first proposed refurbishment (Table 3.6) for the joint between the ceiling and the external wall (detail 2) shown in Fig. 3.5. The results indicate that such a refurbishment would be inappropriate; the result being the temperature of the wood beam would change radically, which would lead to moisture build up and mold growth, and eventually serious structural damage.

To ensure the building's structural integrity, an alternative solution therefore needed to be found. The new option considered involved intentionally leaving a small thermal bridge, where some of the heat from the building could flow out, while keeping the surface temperatures of the wood beam within acceptable boundaries. As part of this assessment, a vapor diffusion analysis was carried out making use of results from the GEQ tool. The results of the thermal bridge and vapor diffusion analysis are shown in Fig. 3.6, while the linear vapor diffusion analysis for over a year is shown in Fig. 3.7. A critical result of this analysis is that when dealing with historical buildings, it is essential to carry out a thermal bridge analysis, so as to avoid potential serious problems, such as those that would potentially arise if the first retrofitting scenario use is employed.

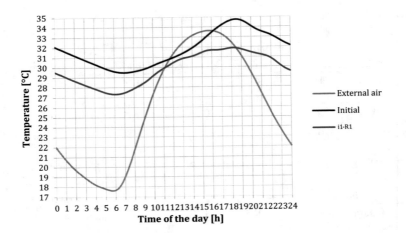

Fig. 3.3 Internal air temperatures in a southwest oriented room in the upper floor for the middle of July, where initial refers to the pre-retrofitting conditions and i1-R1 is the assessed scenario (see Table 3.6)

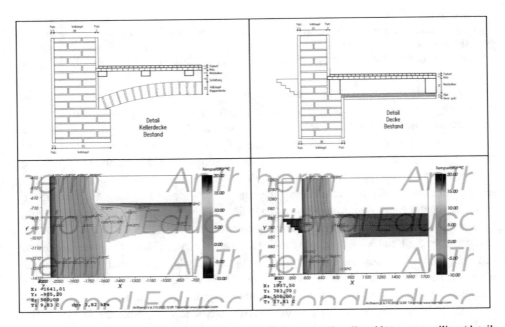

Fig. 3.4 Top: The structure of the joints between (left) the external wall and basement ceiling (detail 1) and (right) the external wall and the ceiling (detail 2). Bottom: The thermal bridge analysis of (left) detail 1 and (right) detail 2

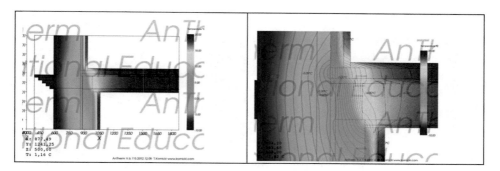

Fig. 3.5 The results of the thermal bridge analysis using the AnTherm software tool for the first proposed refurbishment solution for detail 2. Left: The original thermal bridge results (see Fig. 3.4) and (right) the analysis after the proposed retrofitting

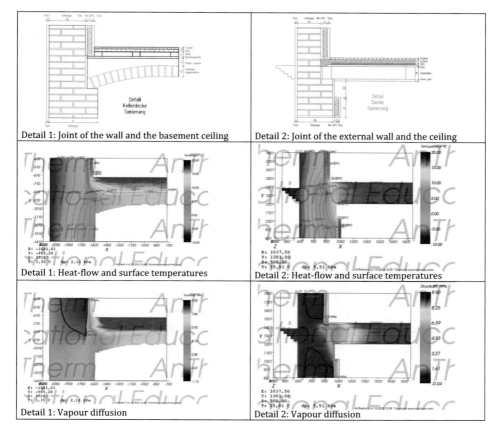

Fig. 3.6 Top: Analyzed refurbishment for detail 1 (left) and detail 2 (right). Middle: The thermal bridge analysis for each detail using results from the AnTherm tool. Lower: The vapor diffusion analysis from the GEQ tool

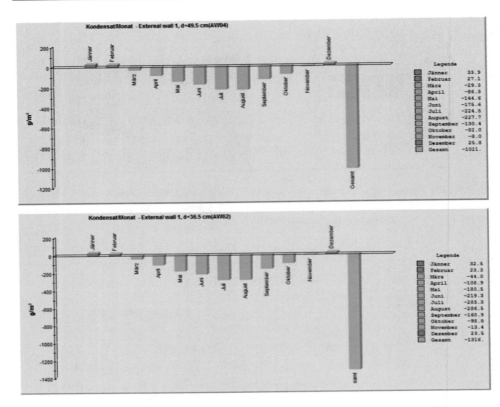

Fig. 3.7 Linear vapor diffusion for (top) detail 1 and (bottom) detail 2 based on the GEQ analysis over 1 year

3.5.2   Case Study 2: Refurbishment of Multistory Housing

The second case study involved evaluating the refurbishment possibilities for multistory residential buildings from the 1970s and 1980s in Serbia (see Fig. 3.8). The target building was located in the city of Belgrade, the capital of Serbia (with 2520 heating degree days). The building had a total of six floors, with a ground floor area of 233.97 m^2 and a total heated floor area of 1403.82 m^2. It had a total volume of 4005.71 m^3, and the façades were oriented toward the south and north. The building's perimeter was 68.7 m, and it had an A/V(external surface area/volume) ratio of 0.4.

Fig. 3.8 A multistory residential building from the 1970s in Belgrade that was the subject of case study 2. The model was positioned within a row of buildings; hence there were only two façades

Table 3.7 Thermal properties of the various elements of the multistory building in Belgrade that was the subject of case study 2

Walls	Cellar ceiling	Roof ceiling	Windows	Frames	Doors
$U_{wall} = 1.68$ W/ m²K	$U_{floor} = 1.16$ W/ m²K	$U_{roof} = 1.73$ W/ m²K	$U_w = 2.91$ W/ m²K (6-16-6 mm)	$U_{frames} = 3.3$ W/ m²K	$U_{doors} = 3$ W/ m²K

The building was of the typical style of construction from the 1970s. The external walls were built using 25 cm thick brick blocks plastered with lime-cement plaster on the inside and decorative "Terranova" plaster on the outside (further details may be found in Jovanović Popović and Ignjatović 2013b). The total window area was 106.88 m², of which 64.19 m² were oriented toward the south. The windows were double glazed, with 80% of their area glass and the other 20% being the wooden frames. The thermal properties of the building's various elements are summarized in Table 3.7.

The refurbishing potential was assessed making use of the indicators listed in Table 3.1, for example, annual heating demands and annual heat gains and losses. The analysis was based on simulations using the Euro-Waebed software in order to evaluate the building's heat energy balance. The simulations were made considering the initial condition of the building and two refurbishment scenarios, termed low-energy and passive house scenarios, each of which is described in Table 3.8.

The low-energy or i2-R1 scenario involved the building's envelope (excluding the cellar) being upgraded by installing 10 cm of EPS insulation, plus new double-glazed krypton-filled windows (4-12-4 mm, i.e., 4 mm glass, 12 mm krypton, 4 mm glass). For the second scenario, the passive house or i2-R2 option, a more comprehensive renovation scheme was proposed. This involved the installation of 20 cm EPS insulation in the envelope, the same double-glazed krypton-filled windows (4-12-4 mm) as for the low-energy option, plus a 75% efficient energy recovery ventilation system. The energy

Table 3.8 The refurbishment options and resulting changes in various energy-use parameters employed for the multistory building in Belgrade that was the subject of case study 2

Scenario	Refurbishment option	Heating demands (kWh/m²a)	Annual heating consumption (kWh/a)	Savings (%)	Life cycle savings (30 years) (kWh)
Initial conditions	–	119	168,078	–	–
Low energy (i2-R1)	Building envelope—10 cm EPS (except basement) Double glazing, 4-12-4 mm	30	42,485	75	3,767,790
Passive house (i2-R2)	Building envelope—20 cm EPS Double glazing, 4-12-4 mm ERV—75% efficiency	2	2671	98	4,962,210

Test i2-R1 is for a low-energy scenario and i2-R2 is for a passive house design

recovery ventilation system is a must-to-have, in order to reach an energy performance of a passive house.

As seen from Table 3.8, the proposed refurbishment measures improved the thermal properties of the building in both scenarios substantially. Considering the heating demands, for the building's original condition, this was 19 kWh/m²a. However, both scenarios saw drastic reductions in this, with savings of around 75% for the low-energy option and 98% when following the passive house scheme.

These results point out how the heating demands for the basic building could be almost totally covered by internal gains (e.g., occupants, appliances) and solar gains. The annual heat losses and heat gains for each scenario were analyzed next, the results being presented in Fig. 3.9.

The findings of this case study clearly demonstrate the huge heat energy saving potential when refurbishing a multistory residential building in Belgrade of the style built in the 1970s and 1980s. The resulting low heating demands after the passive house refurbishment were achievable because in the warmer climate of Southeast Europe, the initial heating demands were lower, in comparison to those in Central Europe (e.g., Vienna). In the following case studies, these two regions will be compared within the context of the refurbishing of detached houses.

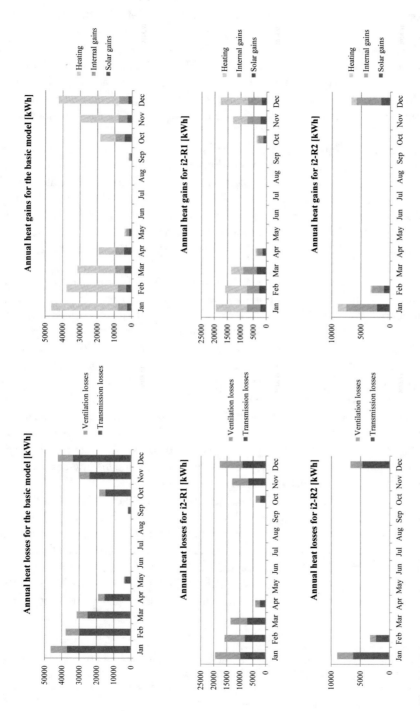

Fig. 3.9 Direct comparison of annual heat losses (left) and annual heat gains (right) for the multistory building used in case study 2. Top: Original conditions, (middle) low-energy option, (bottom) passive house option

3.5.3   Case Study 3: Socioeconomic Influences on Refurbishment

Case study 3 was concerned with the evaluation of the cost-effectiveness of refurbishment, and the influence of local socioeconomic factors in different cities. This will be done by making a comparison between the cities of Niš and Belgrade in Serbia and Vienna (heating degree days of 3459) in Austria. The similar population and area of these two countries, as well as substantial socioeconomic differences, were the key factors in choosing Vienna as the reference point.

The results of the refurbishments are assessed by considering indicators such as the resulting energy saving potential, local energy prices, investment return, and the affordability of the investments. For each location, the same representative house model was considered to allow a precise comparison of the results. The Euro-Waebed simulation software was used to calculate the energy demands for space heating under different refurbishment scenarios. The reason for choosing this parameter is that, as was discussed in the introduction to this book, household heating is the largest consumer of energy in the residential sector in Serbia.

Furthermore, energy prices in Serbia are lower than those in Austria. Interestingly, the pricing of district heating in Serbia up until 2013 was calculated per heated floor area, without considering the actual quantity of energy consumed (district heating system Beogradske Elektrane 2013). Furthermore, there are three categories of electricity pricing in Serbia: (1) green zone, for consumption up to 350 kWh per month; (2) blue zone, for consumption from 350 to 1600 kWh per month; and (3) red zone, for consumption above 1600 kWh per month. Each of these zones in turn has a high-cost tariff for daytime hours (H, 08–24 h) and low-cost tariff during the night (L, 00–08 h) (EPS 2013). Although the market prices for gas, wood, coal, and wood pellets are given per cubic meter or per ton, for a simplified comparison of the heating costs for the three considered cities listed in Table 3.9, the prices are depicted in EURO cent per kWh (€ct/kWh). Note that the price of wood-based heating in Niš and the price of electricity considering the low-cost tariff for the green zone are the cheapest heating sources.

However, there is a trend of increasing energy prices in Serbia (Srbija Gas 2013). In addition to energy prices, another critical socioeconomic parameter related to the affordability of investments into energy saving measures is the average income of a region's inhabitants. For example, average net annual income in Serbia for 2012 was a quarter of that of Austria, with 4536 € per annum in Serbia and 18,529 € in Austria. The difference is even greater when one compares Niš, with 3492 €, and Vienna with 20,594 €, while for Belgrade this is 5100 €.[4] Note that this information was updated in 2013, at the time of conducting this case study.

---

[4]Note, for all economic calculations in this case study, a mid-value of euro exchange rate of the National Bank of Serbia was used (1 EUR = 112.34 RSD, 03.01.2013, National Bank of Serbia 2013, for a comparison, on 15.07.2018 the mid-value was 1 EUR = 118.03 RSD). To obtain realistic results, costs of investments for proposed retrofits were calculated based on an average of at least

Table 3.9 Energy prices by heat sources in Belgrade, Niš, and Vienna

Location	District heating	Gas [€ct/ kWh]	Wood [€ct/ kWh]	Coal [€ct/ kWh]	Wood pellets [€ct/ kWh]	Electricity [€ct/kWh]
Belgrade	11.42 EUR/ m² annually	5.3	3.9	3.9	3.6	Green: $H = 5.3$;
Niš	8.04 EUR/ m² annually	n/a	2.7	3.9	3.6	$L = 1.3$ Blue: $H = 7.9$; $L = 2.0$ Red: $H = 15.8$; $L = 3.9$
Vienna	6.8 €ct/kWh	6.15 + 2.9% p.a.	n/a	n/a	4.57	19.45 + 3.8% p.a.

Note: For the electricity, H indicates the high-cost daytime tariff and L is the low-cost (nighttime) tariff and the color names indicate the different pricing dependent upon the consumption (see main text)

The simulation model employed was a common Serbian two-story family house from the 1970s (Fig. 3.10), while also being common in Austria and Germany. The model was extracted from a database of 1300 typical house designs from this time (Mihailović 1979), with this choice made so as to enlarge the target group who could benefit from the findings. The representative building was defined as being built using a typical concrete skeleton construction with concrete foundations on gravel. The external non-insulated walls were built from brick blocks with mortar finishing. The building had a semi-fabricated ceiling construction with a timber-frame roof, which in turn was made up of two slopes of 40° and covered with clay tiles. The windows were double glazed, made up of 80% glass and 20% wooden frames. Other parameters are listed in

Table 3.10 Finally, the assumptions are made that the house has four occupants, as well as air infiltration, and that it is not surrounded by shading objects, such as trees, neighboring houses, etc.

In the initial situation, the building was without any thermal insulation and had poor thermal properties. The double-glazed windows (6-16-6 mm) had a heat-transfer coefficient of $U = 2.91$ W/m²K. The U-value of the wooden frames was 1.44 W/m²K, while for the wooden doors, it was 2.1 W/m²K.

Four different refurbishment scenarios were evaluated for each of these locations. The proposed measures included the installation of insulation in the external walls and roof ceiling, and the replacement of doors and windows. The first three scenarios involved the installation of differing thicknesses of EPS insulation to all external walls. The first

---

three offers from local companies. The costs of the investments in the cities differed, mainly due to the different costs of working hours.

Fig. 3.10 The simulation
model used in case study 3: a
common two-story family house
from the 1970s. Source: adopted
from Mihailović (1979)

Table 3.10 Input parameters for the example house (Fig. 3.10) that will form the basis for the simulations in case study 3

Feature of the house	Ground floor area (m², heated)	Upper floor area (m², heated)	Façade area (m²)	Total height (m)	Windows area (m²)	Doors area (m²)	Glazing per façade (m²)			
Dimension	76.8	76.8	223.8	9	16.48	2.8	South 6.91	East 1.79	North 1.54	West 1.54

scenario (i3-R1) saw the installation of 5 cm of EPS insulation, a common renovation action done in Serbia over the few decades, termed the "5 cm demit-façade." Such an investment costs 5000 € in Belgrade, 3300 € in Niš, and 9500 € in Vienna. The second retrofit scenario (i3-R2) called for more extensive improvements, involving 10 cm of EPS insulation being installed at a cost of 5750 € in Belgrade, 4500 € in Niš, and 11,000 € in Vienna. The third scenario (i3-R3) set out to find a balance between a more extensive renovation and affordability. In this case, 15 cm of EPS insulation were installed in the external walls, leading to costs of 6150 € in Belgrade, 5200 € in Niš, and 12,500 € in Vienna. The fourth scenario (i3-R4) was a more comprehensive retrofit, which saw the installation of 20 cm of EPS insulation to the external walls. In addition, 10 cm of EPS insulation was installed in the roof ceiling, and the old windows and doors were replaced with new double-glazed windows (4-12-4 mm). This investment resulted in substantially greater costs, with 10,850 € in Belgrade, 9500 € in Niš, and 20,000 € in Vienna.

The applied refurbishment measures were analyzed within the context of potential reduced energy use, in particular, decreased heating needs. The period of the payback period depended on various economic conditions, such as the availability of insulation products on the market, current energy prices, and current socioeconomic parameters such as the ratio of income to investment costs and energy costs themselves.

Table 3.11 Annual heat energy consumption and annual savings for the original and refurbished houses in the three cities: Belgrade, Niš, and Vienna

Cities	Belgrade		Niš		Vienna	
Retrofit scenario for the house (153.2 m²)	Heating demands [kWh/m²a]	Energy savings (%)	Heating demands [kWh/m²a]	Energy savings (%)	Heating demands [kWh/m²a]	Energy savings (%)
Initially	127	–	126	–	160	–
i3-R1	94	26	93	26	120	25
i3-R2	83	35	82	35	107	33
i3-R3	78	39	77	39	100	37
i3-R4	53	58	52	58	70	56

Table 3.12 Investments, annual heating costs by source, and payoff period for refurbishments in Belgrade, Serbia

Retrofit scenarios in Belgrade	Invest (€)	Gas [€/a]	Wood [€/a]	Coal [€/a]	Electr. H-100% [€/a]	Electr. H-67%; L-33% [€/a]	Electr. L-100% [€/a]	Payoff on average, comparison of retrofit scenarios
Initial	–	1060	800	765	2230	1490	565	–
i3-R1	5000	790	600	580	1650	1110	415	22
i3-R2	5750	700	530	510	1430	980	370	19
i3-R3	6150	650	500	470	1290	800	315	18
i3-R4	10,850	460	350	330	770	520	195	20
Payback period, looking at heat source	–	15–19 years	21–25 years	21–27 years	7–9 years	10–14 years	31–36 years	–

The house was simulated for its initial conditions and the four refurbishment scenarios, the results listed in Table 3.11. As can be seen, the percentage savings for all three cities were very similar. Considering the original house, in Belgrade the heating demands were 127 kWh/m²a, while for Vienna it was 160 kWh/m²a, a difference of 33 kWh/m² annually. However, the difference between Belgrade and Niš was negligible. What can be seen from these results is that even with the minimal (first scenario) renovation, savings of around 25% can be achieved. These findings point out that even with the minimal renovation, 1/4 of energy can be saved, while considering the most extensive renovations (fourth scenario) sees the houses in Serbia improved to the low-energy standard (52 kWh/m²a).

To evaluate the cost-effectiveness of the refurbishments, the annual costs of heating were calculated, which, as one would expect, differed greatly for each case, owing to the different prices of the used energy source. For Belgrade, the annual costs of district heating were fixed to 1750 € annually, regardless of the reduced energy consumption. This disparity in heating costs resulted in a range of possible payoff periods (Table 3.12). The shortest average payoff period was for the third scenario, since the biggest savings are achieved when electricity was used in the high tariff period for heating, which under such conditions saw the investment paid off in 7 years.

The Niš case was found to be the least expensive, as were the annual heating costs. The district heating was fixed to 1225 € annually, regardless the energy performance of the building. Despite the fact that the energy prices were the cheapest of the examples considered, this case showed the quickest financial return (Table 3.13). As in the Belgrade case, the third refurbishment scenario showed the quickest average payoff period. As electricity prices for the high-cost tariff was the most expensive heating source, the payoff period for this scenario in the case of electricity (100% of consumption during the high-cost tariff period) was only 6 years.

Finally for the Vienna case, the required investments were the largest, around double that in the Serbian cities, essentially because of the higher labor costs. However, the financial return again also depended on the energy source (Table 3.14). For example, the cost-effectiveness of refurbishment was greatest for the case of electrical heaters (initial heating cost of 4950 €/a), with the quickest payback period again for the third scenario with an average period of 20 years. This sees the payoff period for electricity within the third scenario being only 7 years.

Although the i3-R3 option demonstrated the quickest return on any investment, scenario i3-R4 showed an additional 19% in energy savings (Table 3.11). Considering a period of 30 years for Belgrade and Niš, employing the i3-R4 refurbishment scenario would save up to 112,000 kWh more than the i3-R3 option. Hence, the i3-R4 option was, over the long term, the most profitable solution, although under current energy prices, the financial difference was minimal.

An unexpected result was that financial savings were also possible when not undertaking any refurbishment owing to the disparity in the prices associated with the different energy sources. This means that by simply changing the source of energy, significant financial savings could be made. Considering the original non-insulated house in Belgrade, changing the electricity tariff from high-tariff to low-tariff alone (e.g., the use of heating equipment that consume electricity during the night to heat up the thermal storage, and then release the heat energy during the day) achieved greater financial benefits than from the comprehensive retrofit. The same situation was found in Niš, where changing from the high to low electricity tariff, or to wood heat, saved about 1650 € per heating season.

Naturally, affordability of any investment depends on the income of the house owner, a major socioeconomic parameter when considering refurbishing. For instance, for Belgrade, the cost of the investment for the i3-R3 scenario was equal to 15 average monthly salaries, in Niš this was 18 and for Vienna 8, all owing to the average incomes as outlined above. The case is even more severe when we consider the highest-level intervention (i3-R4), where in Belgrade this is equal to 26 average monthly salaries, 33 in Niš, while in Vienna 12 were needed (Table 3.15).

Moreover, in Fig 3.11, relationship between the annual heating costs of non-refurbished house, the average investment, and annual income (i.e., the average of the four scenarios considered in case study 3) required refurbishment in the three assessed cities. In case of

Table 3.13 Investments, annual heating costs by source, and payback period for refurbishment in Niš, Serbia

Retrofit scenarios in Niš	Invest (€)	Wood [€/a]	Coal [€/a]	Electr. H-100% [€/a]	Electr. H-67%; L-33% [€/a]	Electr. L-100% [€/a]	Payoff on average, comparison of retrofit scenarios
Initial	–	550	760	2210	1480	560	–
i3-R1	3300	420	560	1630	1100	410	17
i3-R2	4500	370	500	1440	970	365	17
i3-R3	5200	350	460	1270	790	310	16
i3-R4	9500	250	320	760	510	190	18
Payback period, looking at heat source	–	25–32 years	17–22 years	6–7 years	8–9 years	22–26 years	–

Table 3.14 Investments, annual heating costs by source, and payback period for refurbishments in Vienna, Austria

Retrofit scenarios in Vienna	Invest (€)	District heating [€/a]	Gas [€/a]	Pellets [€/a]	Electricity [€/a]	Payoff on average, comparison of retrofit scenarios
Initially	–	1665	1550	1120	4950	–
i3-R1	9500	1250	1170	880	3710	24
i3-R2	1100	1120	1040	750	3310	20
i3-R3	12,500	1050	970	700	3100	20
i3-R4	20,000	730	680	500	2170	22
Payback period, looking at heat source	–	21–23 years	22–25 Years	30–40 years	7–10 years	–

Table 3.15 The costs and the number of average monthly salaries that would equal the required investment for each refurbishment option

Retrofit scenario	Belgrade		Niš		Vienna	
	Investment (€)	Number of average monthly salaries equal to investment	Investment (€)	Number of average monthly salaries equal to investment	Investment (€)	Number of average monthly salaries equal to investment
i3-R1	5000	12×	3300	12×	9500	6×
i3-R2	5750	14×	4500	16×	11,000	7×
i3-R3	6150	15×	5200	18×	12,500	8×
i3-R4	10,850	26×	9500	33×	20,000	12×

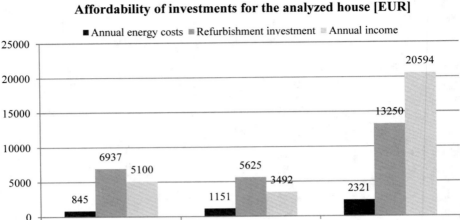

Affordability of investments for the analyzed house [EUR]

Fig. 3.11 Relationship between the annual heating costs of non-refurbished house, the average investment, and annual income (i.e., the average of the four scenarios considered in case study 3) required refurbishment in the three assessed cities

Belgrade and Niš, the average refurbishment investments overlapped the annual income, while for Vienna, the average refurbishment investment was lower than the average income.

3.5.4   Case Study 4: Refurbishment According to Energy Policies

The next case study sets out to examine and evaluate policy-driven aspects of refurbishments. The clarification of legal requirements is important when trying to encourage house owners to invest in retrofitting, especially since energy efficiency policy in Serbia has changed a great deal over recent years. For example, the Rulebook on Energy Efficiency in Buildings from 2011 outlined a set of criteria dealing with building energy performance. Afterward, three NEEAPs were released. Similarly, the first NEEAP proposed suggestions to improve building energy efficiency by 9% by 2018, while the second NEEAP was published within the year 2014 and the third NEEAP in 2016.

Therefore, this case study is based on the following research questions: (1) what specific refurbishment measures meet the targets of current and future energy efficiency policy in Serbia; and (2) based on these policies, what is the energy saving potential, improvements in thermal comfort, and the economic effect of such refurbishments? Hence, one sees that this assessment considered environmental, social, and economic characteristics. The case study again, as in case study 2, involves the analysis of a typical Serbian house design from the 1970s and 1980s located in Belgrade. However, this time the subject is a two-story

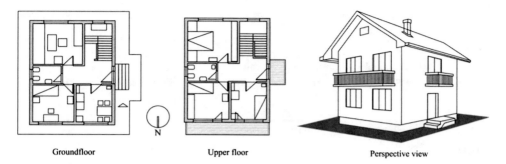

Groundfloor   Upper floor   Perspective view

Fig. 3.12 The representative two-story family house used as the simulation model for case study 4 (this building is denoted by the code "P+1-2/1" in the catalogue of typical housing designs. Source: adopted from Mihailović (1979)

Table 3.16 Properties of the building envelope for the representative two-story single-family house selected for the simulation studies as part of case study 4

Building element	Ground floor	External walls	Roof ceiling	Windows	Frames	Doors
Surface	63 m^2	206.7 m^2	63 m^2	23.12 m^2	4.62 m^2	3 m^2
Thickness	56.6 cm	22 cm	21.5 cm	6-12-6 mm	22 cm	4 cm
U-Value [W/m^{2}k]	1.52	0.8	0.89	2.91	1.44	2.1

single-family house. The characteristics of the buildings selected for assessment are from the catalogue of typical housing designs of the former Yugoslavia from this period (Fig. 3.12, Mihailović 1979).The specific house design chosen was designed by Milisav Vojinović, the architect of several buildings in the catalogue.

The example house has a concrete skeleton structure whose external walls are constructed of brick blocks with plaster finishing and no thermal insulation. Its total floor area was 126 m^2, with the ground and upper floors being heated zones, while the attic was not heated. The building had a semi-fabricated ceiling and timber-framed roof made up of two slopes of 30°. The total height of the house was 8.7 m, with the south-oriented windows having a total surface area of 7.32 m^2. The windows were double glazed with 80% of the area made up of glass and 20% by wooden frames. The thermal properties of the building's elements are listed in Table 3.16.

For the dynamic simulations, the Euro-Waebed software took into account the presence of the occupants, heat gains from equipment, as well as air infiltration, while assuming the building was not shaded. For the heating demands evaluation, the indoor temperature during the heating period was set to 20 °C, as outlined by Zrnić and Ćulum (1998). The GEBA simulations considered the south-oriented living room on the ground floor as the potentially overheated space, with the analysis carried out for the middle of July.

Table 3.17 The designed scenarios for refurbishment as part of case study 4, as defined by different policies

Scenario	According to	Target
i4-R1	2011 Rulebook	Boundary U-values
i4-R2	1st NEEAP	Reduce for 50 kWh/m^{2}a comparing to initial (non-retrofitted) condition
i4-R3	2nd NEEAP	Reduce to 75 kWh/m^{2}a
i4-R4	Passive house standard	Reduce to 15 kWh/m^{2}a

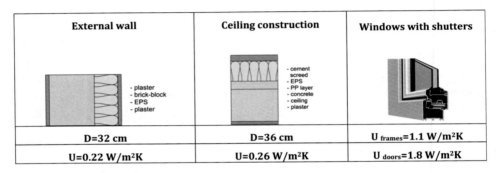

External wall	**Ceiling construction**	**Windows with shutters**
- plaster - brick-block - EPS - plaster	- cement screed - EPS - PP layer - concrete - ceiling - plaster	
D=32 cm	**D=36 cm**	**U frames=1.1 W/m²K**
U=0.22 W/m²K	**U=0.26 W/m²K**	**U doors=1.8 W/m²K**

Fig. 3.13 Details of the building elements used for the third retrofit scenario (i4-R3) in case study 4

Four refurbishment scenarios were considered, as summarized in Table 3.17, each a function of a different energy efficiency policy. The first option (i4-R1) was developed according to the 2011 Rulebook on Energy Efficiency in Buildings. This set boundary conditions on the thermal properties for new and refurbished building elements. The external walls and roof ceiling were retrofitted by the installation of 6 cm EPS insulation to achieve a heat-transfer coefficient of less than 0.40 W/m^{2}K. The second option (i4-R2) was designed with respect to the first NEEAP targets which set out to reduce the heating demands of houses by at least 50 kWh/m^{2}a (Todorović 2010). In this case, the walls and roof ceiling were modified by the installation of 8 cm of EPS insulation. The third refurbishment option (i4-R3) was designed in relation to the second Serbian NEEAP and EPBD directive 2010/31/EC and called for the reduction of heating demands for refurbished buildings to 75 kWh/m^{2}a (Batas Bjelic et al. 2013). This employed the upgrading of the walls and the roof ceiling by installing 10 cm of EPS and replacing existing windows with shuttered double-glazed glass (see Fig. 3.13). The fourth scenario (i4-R4) set out to attain national passive house standards with an energy consumption of 15 kWh/m^{2}a for the net floor area. This required the walls, basement ceiling, roof ceiling, and roof skin to be insulated with thick layers of EPS (14–25 cm) and extruded polystyrene

Table 3.18 Financial return on the investment into the four refurbishment scenarios for case study 4

Scenario	Total investment (€)	Investment (€) per m^2	Savings in 30 years (MWh)	Payoff (years)
i4-R1	4200	33	177	15
i4-R2	5000	40	204	15
i4-R3	8800	70	278	20
i4-R4	28,300	225	491	35

foam insulation (10 cm). In addition, the windows and doors were replaced with units offering better thermal properties, superior airtightness, and improved U-values. In addition, a 75% efficient heat recovery ventilation was installed, and the thermal bridge impact was reduced. Such refurbishment measures could in fact be reapplied on typical single-family houses to meet the current and possible future policies. The results of these retrofitting efforts are reported by presenting the modeled annual heating energy consumption, the internal air temperatures in summer, and the investment return for each refurbishment scenario. These, as indicated above, cover environmental, social, and economic indicators.

For the environmental (energy usage) assessment, the case study showed that the heating demands for the basic house were 145 kWh/m^{2}a. The resulting reduction on the heating demands due to the refurbishments ranged from 32% for option i4-R1, 37% for i4-R2, 52% for i4-R3, and an impressive 90% for i4-R4. It should also be commented that the proposed refurbishment scenarios reduced the overall environmental impact of the building, since carbon dioxide emissions from energy production would be reduced.

From an economic standpoint, the partial renovations following i4-R1 and i4-R2 were the most affordable options. The i4-R3 scenario showed a better ratio of investment return than i4-R4, which as expected involved the highest initial investment. Over 30 years, it can be seen that the refurbishments allowed significant energy savings (Table 3.18). However, because of the relatively low cost of energy, the payoff periods were relatively long, ranging from 15 to 35 years. The investment return assessment was calculated considering the average price of gas, wood, and electricity, which was 4.9 ct/kWh, in the case of a cash investment. For the situation where a renovation loan or "energy efficiency loan" was taken, an example of such an offer was provided by the "ProCredit Bank" (15.04.2013, ProCredit Bank 2013). Considering a repayment period of 5 years and an effective interest rate of 8.35%, this would extend the payoff period by an additional 6 years for the i4-R1 scenario, 8 years for i4-R2, and over 10 years for i4-R3 and i4-R4. It needs to be said that until 15.04.2013, renovation loans in Serbia were not favorable in comparison with other EU countries (e.g., Austria and Germany). This therefore underlines one of the economic obstacles to refurbishing. Note that this and the previous case study were dependent upon specific offers from construction firms, obtained at the time, hence potentially subject to change.

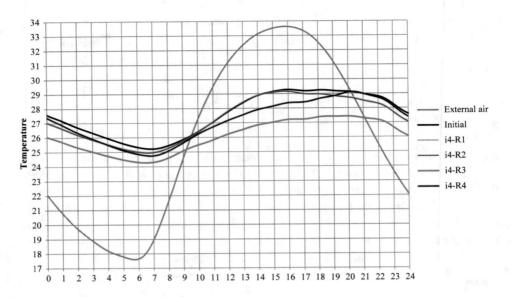

Fig. 3.14 Air temperatures during the day in south-oriented living room on 15th of July for the unmodified house and for each refurbishment scenario compared with outside temperatures. Note that the results for i4-R1 and i4-R2 are almost the same, such that i4-R2 overlaps i4-R1 in this figure

Considering now the social aspect of the refurbishment, namely, the summer thermal comfort, the air temperature inside the south-oriented room for the unmodified house and that resulting from the four options are shown in Fig. 3.14. For the i4-R1 and the i4-R2 scenarios, a small decrease in the inside temperatures is seen with respect to the unmodified house. The i4-R3 option provided the best result under critical summer conditions, with a maximum temperature decrease of about 2 °C. However, the i4-R4 scenario showed overheating potential and required additional cooling to achieve optimal thermal comfort in the room. This is due to the triple-glazed south-oriented windows creating a kind of "glasshouse" effect due to the high air temperatures in summer months. It therefore demonstrates how care must be taken when applying typical passive house principles in Southeast European countries.

These environmental, economic, and social assessments within the context of scenarios based on various energy policies provide a good basis for the further development of systematic approaches to the design of appropriate refurbishment strategies for specific types of buildings.

3.5.5   Case Study 5: Design Patterns for House Refurbishment

The aim of this case study is to assess the effectiveness of individual refurbishment measures for common detached houses in Serbia and to develop a design matrix of feasible

Two-storey house
"P+1-6" in Belgrade

Two-storey house
"P+1-2/1" in Nis

Two-storey house with basement
"HP+1-116" in Belgrade

Fig. 3.15 The three house models extracted from the typical housing designs catalogue. Source: adopted from Mihailović (1979)

design solutions or design patterns. Three types of houses were selected to be the subject of the simulations, denoted by the catalogue of typical housing designs from the 1970s and 1980s in Serbia and other ex-Yugoslavian Republics as "P+1-6," "P+1-2/1" (which was also the subject of the previous case study, see Fig. 3.12), and "HP+1-116" (see Fig. 3.15). The objectives of this case study may therefore be summarized as (1) the evaluation of the effectiveness of specific parameters employed as energy conservation measures and (2) to present the results in the form of patterns or a toolbox that could be used to support the process of designing the refurbishment of a building. The analyzed locations are again Belgrade and Niš.

The methodology employed for this case study followed Konstantinou's method for the refurbishment of multistory residential buildings (Konstantinou 2014). Since, as has been mentioned, heating demands make up the largest proportion of the energy consumption in Serbian buildings, this parameter is used as an energy efficiency indicator (Todorović 2010). To determine this parameter for the houses and then evaluate the effectiveness of the individual retrofitting measures, the tool Euro-Waebed was used. The presence of occupants was taken into account, as was heat gains from equipment and air infiltration. For the dynamic simulations, the indoor temperature during the heating period was set to 20° C.

The simulations involved changing only one building element at a time relative to the reference building model for each of the selected building types. This enabled the contribution of each measure to be evaluated. There were various scenarios for the walls, including options with EPS for external insulation and one option with capillary thermal insulation for internal application. The ground floors were insulated using extruded polystyrene foam (XPS). Replacing the windows involved considering both double and triple glazing, while airtightness was dealt with by sealing any air leaks around the windows. Included in the roof options are green roofs, which are roofs that are flat with a layer of soil and grass grown on them. For ventilation, a heat recovery device with a 95% efficiency was employed. All of the renovation options were thus systematically arranged in a form of a "toolbox," whose various parameters are listed in Table 3.19

Table 3.19 The toolbox of energy conservation measures employed in case study 5

External walls	Floor	Roof	Windows	Thermal bridges	Air supply
No insulation	No basement	Roof ceiling—not insulated	Single glazing	Linear thermal bridges (e.g., balconies)	Window ventilation
Little/ outdated insulation	Floor on the ground—not insulated	Roof ceiling—insulated	Double uncoated	Geometrical thermal bridges (e.g., junctions)	Ventilation with heat recovery
External insulation ETICS 10 cm	Floor on the ground—insulated	Pitched roof—not insulated	Upgrade existing windows	Repeating thermal bridges (e.g., joists)	
ETICS advanced (EPS 20 cm)	Basement ceiling—not insulated	Pitched roof—insulated	Replacement: 2× glazing	Thermal bridges—partly insulated	
Internal insulation	Basement ceiling—insulated below slab	Green roof	Replacement: 3× glazing	Thermal—bridge-free construction	
Ventilated façade			Shading		

The resulting toolbox was used to diagnose the initial conditions, as well as developing two refurbishment patterns by selecting a set of measures for each specific case. For the unmodified buildings, the houses were concrete structures with brick block walls without insulation. The U-value of the non-insulated external walls was 0.80 W/m²K, while the windows had low performance, and the envelope was not airtight.

For the first pattern, refurbishment scenario (i5-R1), the intention was to achieve a 50% saving in heating demands. Hence, 10 cm of EPS insulation was installed on the walls and the roof ceiling, along with the replacement of the widows with double-glazed glass. For the second pattern, refurbishment scenario (i5-R2), the intention was to attain passive house standards, requiring a reduction in the energy demands by a factor of 10. This required the installation of 20 cm of EPS insulation in the walls and ceilings, while the basement slab and the roof skin were retrofitted with 10 cm of EPS insulation. The airtightness was improved, the windows replaced by double-glazed windows with shutters, and a 95% efficient heat recovery ventilation system installed. The two scenarios are presented in the toolbox outlined in Fig. 3.16.The upgrading of the buildings were intended to improve their thermal properties, to reduce the overall heat-transfer coefficient (U-value), to replace openings (i.e., windows) with higher quality components, and to install high-energy recovery mechanical ventilation.

(left) initial models

WALL	FLOOR	ROOF	WINDOW	AIR
No insulation	No basement	Roof ceiling- no ins.	Single glazing	Window ventilation
Outdated insulation	Ground-floor- no ins.	Roof ceiling- insulated	Double uncoated	Non – airtight envelope
ETICS standard 10	Ground-floor – insulated	Pitched roof – no ins.	Upgrade of existing windows	Airtight envelope
ETICS advanced 20	Basement ceiling – no ins.	Pitched roof – insulated	Replace: 2x glazing	Ventilation with heat recovery
Internal insulation	Basement ceiling – insulated	Green roof	Replace: 3x glazing	
Ventil. Facade			Shading	

(center) scenario i5-R1

WALL	FLOOR	ROOF	WINDOW	AIR
No insulation	No basement	Roof ceiling- no ins.	Single glazing	Window ventilation
Outdated insulation	Ground-floor- no ins.	Roof ceiling- insulated	Double uncoated	Non – airtight envelope
ETICS standard 10	Ground-floor – insulated	Pitched roof – no ins.	Upgrade of existing windows	Airtight envelope
ETICS advanced 20	Basement ceiling – no ins.	Pitched roof – insulated	Replace: 2x glazing	Ventilation with heat recovery
Internal insulation	Basement ceiling – insulated	Green roof	Replace: 3x glazing	
Ventil. Facade			Shading	

(right) scenario i5-R2

WALL	FLOOR	ROOF	WINDOW	AIR
No insulation	No basement	Roof ceiling- no ins.	Single glazing	Window ventilation
Outdated insulation	Ground-floor- no ins.	Roof ceiling- insulated	Double uncoated	Non – airtight envelope
ETICS standard 10	Ground-floor – insulated	Pitched roof- no ins.	Upgrade of existing windows	Airtight envelope
ETICS advanced 20	Basement ceiling – no ins.	Pitched roof- insulated	Replace: 2x glazing	Ventilation with heat recovery
Internal insulation	Basement ceiling- insulated	Green roof	Replace: 3x glazing	
Ventil. Facade			Shading	

Fig. 3.16 Toolbox for the diagnosis of the initial conditions and the i5-R1 and i5-R2 design patterns considering (left) initial models, (center) scenario i5-R1, and (right) scenario i5-R2

An assessment of the results will allow the effectiveness of the individual measures for typical Serbian houses to be identified, allowing a simple means of selecting options for house renovations. The toolbox provides a database of refurbishment potential and highlights the most effective building envelope retrofitting measures. The more detailed description of this process outlined below will demonstrate how selecting the most appropriate measures from the toolbox could enrich the process of developing the required refurbishment scheme. This pattern-refurbishment approach will also allow the assessment of the early design decisions on the overall energy efficiency of the final building.

The analyzed house types differed in their energy demands, with the heating demands for the unmodified houses varying from 107 kWh/m^{2}a for type HP+1-116, 130 kWh/m^{2}a for type P+1-6, and 153 kWh/m^{2}afor P+1-211 (see Table 3.20). This is the result of the differing numbers of floors, the form of the building, and its volume. Although the applied measures showed a similar level of effectiveness, the "P+1-6" model showed slightly better results due to the reduced thermal bridging (i.e., concrete balconies) in its construction.

The energy saving potential of the considered measures are presented in Table 3.20. The installation of insulation in the external walls showed considerable energy saving potential, with this measure alone contributing a saving of up to 40%. Installing floor insulation produced a saving of 11%, comparable to the 8% gained by basement ceiling insulation (note, insulating a basement slab is simpler to undertake). The roof ceiling insulation produced up to 15% of the savings. It is interesting to note that there was an equal reduction in energy demands using either double- or triple-glazed windows. Finally, the ventilation system making use of energy recover ventilators (ER) reduced the initial energy consumption by an average of 15%.

The saving potential of the individual measures is also shown for easier understanding in Fig. 3.17. It is important to note how each measure that involved insulating the external walls resulted in greater energy savings than any of the others, indicating that walls are the elements that need to be considered in any refurbishment process. It is also obvious how less efficient the other measures (generally resulting in a third or less of the savings when compared to the wall insulation), in particular insulating the basement floor (1%) and ceiling (8%) and using a green roof (5%).

Considering the results of the toolbox's application to house type "HP+1-116" (Fig. 3.15) in more detail, as commented above, initial energy consumption of the house was 107 kWh/m^{2}a (see Table 3.20). Employing the first retrofitting scenario (i5-R1) saw energy consumption being reduced to 45 and 11 kWh/m^{2}a for the second scenario (i5-R2), the latter meeting the requirements for passive house energy efficiency. Furthermore, heating demands were reduced by 58% and 90% by the application of scenarios i5-R1 and i5-R2, respectively.

Table 3.20 Energy savings for each of the test house types by the use of individual measures (see also Fig. 3.17)

Building feature	Retrofit measure	U-value [W/m^{2}K]	P+1-6 130 kWh/ m^{2}a (%)	P+1-2/1 153 kWh/ m^{2}a (%)	HP+1-116 107 kWh/ m^{2}a (%)	Average effectiveness (%)
Ext. walls	External insulation 10 cm	0.26	34	32	32	33
	External insulation 20 cm	0.16	40	38	38	39
	Internal insulation 10 cm	0.263	32	30	27	30
	Ventilated façade (10 cm ins.)	0.26	34	32	32	33
Floor	Ground-floor on the ground 10 cm	0.33	11	11	–	11
	Basement floor 10 cm	0.33	–	–	<1	1
	Basement ceiling 10 cm	0.26	–	–	8	8
Roof	Roof ceiling 10 cm	0.26	13	10	11	11
	Roof ceiling 20 cm	0.16	15	12	13	13
	Roof skin (10–14 cm)	0.30	7	6	7	7
	Green roof	0.55	5	5	5	5
Windows	Double glazing (4-12-4 mm, Kr)	1.10	7	12	15	11
	Triple glazing (4-8-4-8-4 mm, Kr)	0.70	7	12	15	11
Air supply	Ventilation with ERV 95%	–	16	12	17	15

Note: The blank spaces are for those measures not applicable for the house (i.e., that measure was not possible)

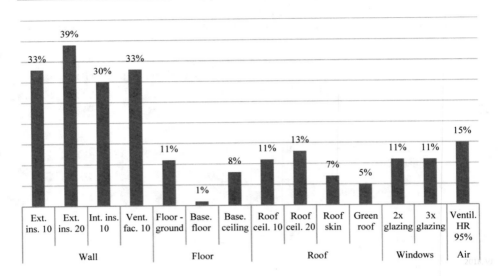

Fig. 3.17 Average energy saving potential of the individual measures (see Table 3.20)

3.5.6   Case Study 6: Replacing a Heating System

From the above case study, the environmental impact of buildings resulting from their consumption of energy may be significantly reduced through the sorts of refurbishments scenarios presented above. In addition to the building's envelope renovations, which saw significant savings in energy resulting from even relatively simple steps, further savings could be made by considering more efficient heating systems.

As commented at the start of this chapter, effort must also be expended in considering energy conversion efficiency and the CO_2 footprint of different heat sources in Serbia. Therefore, this case study has the aim of assessing the energy saving potential and environmental impact of replacing an old inefficient heating system in a household in Serbia. The calculations are done for a single-family house with 150 m^2 heated floor area, with an annual heating demand of 20,000 kWh. The unmodified house makes use of a wood stove for heat generation. The reasons for selecting a wood stove as the heat source include the fact that a large number of households in Serbia still use wood as their source of heating, with 36% of households being heated by wood stoves and 43% of all heat energy being produced from wood (Energy Agency Serbia 2011). To emphasize this point, Fig. 3.18 presents the distribution of heat energy sources in Serbian households and the total heat energy consumption by source. The second reason for choosing the case of wood stoves is that solid fuel furnaces for households have a rather low-energy conversion factor, showing η values varying from 0.65 to 0.68. Finally, wood can be used as a raw material for more enduring purposes other than space heating.

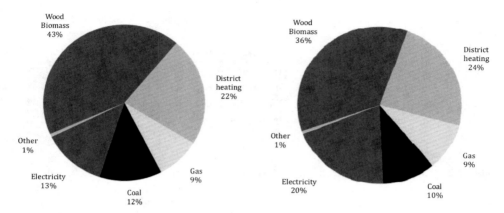

Fig. 3.18 Left: The share of the total consumption by households of the different sources of energy as heat sources. Right: The proportion of households that use each source of energy. Source: Energy Agency Serbia (2011)

In the unmodified case, the house used a wood stove for generating heat energy, the pipe grid used for the heat distribution was not insulated, and the whole building was one thermal zone. The energy conversion efficiencies for this case were, therefore, for generation $\eta k = 0.65$, distribution $\eta c = 0.95$, and in delivery $\eta r = 0.9$, leading to a total energy conversion efficiency of $\eta = 0.55$. This lead to a final energy demand for the reference case, which has a heating demand requirement of 20,000 kWh (see above) of 36,400 kWh, or 182% of the useful energy.

Four scenarios proposing different schemes for replacing the wood stove heating system were considered (see Table 3.21). The replacements considered were a gas boiler (option i6-R1), a wood pellet stove (i6-R2), connecting the house to a district heating system (i6-R3), and an air source heat pump (i6-R4). Each of these options had higher-energy conversion efficiency (or coefficient of performance or COP in case of the heat pump), especially in the generation phase. The evaluating criteria were the reduction with respect to the reference case of the initial final energy for predefined heating demands and the CO_2 footprint. For the CO_2 footprint, the sum of the annual final energy for heating purposes was multiplied by the appropriate CO_2 conversion factor.

Table 3.22 lists these conversion factors for the different heat sources, including coal and electricity, although these were not considered in this case study. The reason for this is that coal has the same energy conversion efficiency as wood and the efficiency of electric space heating is almost 100%.

The results, presented in Table 3.23, demonstrate the energy saving potential of replacing inefficient heating equipment, as well as reducing CO_2 emissions. To deliver the required useful energy, the house with a wood stove required an additional 82% of the final energy, being seen as the least efficient heating solution, with gas, wood pellets, district heating, and the heat pump all displaying lower final energy demands. This shows

Table 3.21 The four scenario options for replacing the heating system of the example house for case study 6

Scenario	Production	Distribution	Delivering
i6-R1	Gas boiler, $\eta k = 0.84$	Insulated grid, $\eta c = 0.98$	Zoning, $\eta r = 0.98$
i6-R2	Pellet stove, $\eta k = 0.90$	Insulated grid, $\eta c = 0.98$	Zoning, $\eta r = 0.98$
i6-R3	District heating (DH) system	Pre-insulated DH grid, $\eta c = 0.90$	Zoning, $\eta r = 0.98$
i6-R4	Heat pump (air), $COP = 3.80$	Insulated grid, $\eta c = 0.98$	Zoning, $\eta r = 0.98$

Table 3.22 CO_2 emissions for a house of 150 m^2 floor area and 20,000 kWh of annual heat consumption

Heat source	CO2 emission per kWh (kg/kWh)
Wood furnace	0.015
Coal furnace	0.32
Gas boiler	0.20
Pellet stove	0.015
District heating system	0.33
Heat pump (air; electricity)	0.53
Electric heaters	0.53

Source: Rulebook on Energy Efficiency in Buildings (2011) and OIB (2011)

Table 3.23 The results obtained from the four scenarios dealing with replacing the wood heating system of the example house

Scenario	Production	Distribution	Delivering	Total efficiency	Final energy (%)	Energy losses (%)
i6-R1	Gas boiler $\eta k = 0.84$	Insulated grid $\eta c = 0.98$	Zoning $\eta r = 0.98$	$\eta = 0.81$	123	23
i6-R2	Pellet stove $\eta k = 0.90$	Insulated grid $\eta c = 0.98$	Zoning $\eta r = 0.98$	$\eta = 0.86$	116	16
i6-R3	District (DH) heating system	Pre-insulated DH grid $\eta c = 0.90$	Zoning $\eta r = 0.98$	$\eta = 0.87$	115	15
i6-R4	Heat pump (air) $COP = 3.80$	Insulated grid $\eta c = 0.98$	Zoning $\eta r = 0.98$	COP 3.54	28	–

Table 3.24 CO_2 emissions for a house of 150 m^2 floor area and 20,000 kWh of annual heating energy consumption for different sources of heat energy

Heat source	Final energy demands (kWh)	CO_2 emission conversion factors (kg/kWh)	Annual CO_2 emission (t)
Wood furnace	36,400	0.015	0.546
Coal furnace	36,400	0.32	11.648
Gas boiler	24,600	0.20	4.92
Pellet stove	23,200	0.015	0.348
District heating system	23,000	0.33	7.59
Heat pump (air source)	5600	0.53	2.968
Electric heaters	20,000	0.53	10.6

Note: Wood is the least efficient, and the heat pump is the most

that the most common heat energy source in Serbia seems to be, at the same time, the least efficient.

Moreover, the proposed heating systems were evaluated by the resulting CO_2 footprint (again including coal and electricity). In this comparison (Table 3.24), we see wood and wood pellets are the most environmentally friendly sources, despite having low-energy conversion efficiency, while coal and electricity (which are often used) are seen to have a more negative environmental impact. These results, along with the energy efficiency ones, illustrate the difficulties that will be encountered when formulating policies for more efficient energy usage and reduced CO_2 emissions.

3.6   Summary

The work presented in this chapter considered the various issues surrounding the refurbishing of the building stock of Serbia, including both historical and more recent (1970s to 1980s) residential buildings. A series of case studies were carried out, targeting different types of residential buildings (historical, multistory, and single family), and issues surround socioeconomic factors, policies, the development of a design matric for different design patters, and sources of energy and CO_2 reduction.

A number of scenarios that consider different ways of refurbishing were evaluated. The most efficient measures involved upgrading the buildings' envelope by different amounts of insulation. However, it was seen that each refurbishment project is unique and needs to be planned and executed with a great attention. This was particularly noted when considering the level of window glazing, which in some cases leads to negative effects in terms of overheating a residence's interior.

With regard to the toolbox developed within this chapter, a set of effective refurbishing measures were proposed and the accompanying patterns developed, which, depending on the early design decisions made, may lead to energy savings of the order of 90%. Nevertheless, some socioeconomic and financial factors, such as the length of time required to pay off the initial refurbishing investment by energy savings and the actual relatively high cost of the initial investment compared to the house owner's income, need to be considered carefully, as they may prove to be obstacles to retrofitting.

The energy saving potential of replacing inefficient heating equipment was also investigated, along with the differences arising in CO_2 emissions when considering the different energy and heat sources available in Serbia. The results showed that there is a great energy saving potential in the replacing of non-efficient heating devices. Moreover, environmental impact of different heating sources differed significantly, and both factors, efficiency and level of emissions, need to be considered together.

The next chapter will consider future building designs, in part making use of the design toolbox developed within this chapter. A series of seven assessment criteria will be employed in the different design options. As in this chapter, case studies (four) will make use of the simulation tools discussed above. Factors such as the geographical dependence and different types of housing will be examined, as well as solar energy strategies.

References

AnTherm. (2014). AnTherm (Thermal Heat Bridges), retrieved from http://antherm.eu/antherm/EN/, (accessed 10.03.2014).

Austrian Energy Agency. (2013). Total energy consumption by sector - Serbia 2010. http://www.enercee.net/countries/country-selection/serbia.html (accessed 10.04.2013).

Batas Bjelic et al. (2013). Improvements of Serbian-NEEAP based on Analysis of Residential Energy Demand until 2030, In Proceedings, 8th IEWT conference, Vienna, Austria.

Belgrade Chamber of Commerce. (2010). Sectoral Collaboration Project with Regard to Financing Energy Efficiency in Buildings within the Frame of EU Regulations and Legal Arrangements, Country report: Serbia. EUbuild Energy Efficiency, IPA Project, country report Serbia. Retrieved from http://www.eubuild.com/wp-content/uploads/2011/06/9CountryReport-SERBIA1.pdf (accessed 10.02.2013).

Bointner R. et al. (2012): Gebäude maximaler Energieeffizienz mit integrierter erneuerbarer Energieerschließung, Berichte aus Energie-und Umweltforschung 56a/2012. Bundesministeriums für Verkehr, Innovation und Technologie, Wien, 2012.

District heating system Beogradske Elektrane. (2013). Belgrade.

Energy Agency Serbia. (2011). Regulation of electricity prices, retrieved from http://www.aers.rs/FILES/Prezentacije/2012-12-3%20EPS%20RegulCena%20dec%202012%20LJM%20c.pdf, (accessed 10.12.2012).

EPS. (2013). Electric power industry of Serbia, http://www.eps.rs/

GEQ. (2014). Dampfdiffusion, retrieved from http://www.geq.at/pdf/hb_dampf.pdf (accessed 10.12.2014).

Hulscher, W. S. (1991). Basic energy concepts. Training Materials for Agricultural Planning (FAO), retrieved from http://www.fao.org/docrep/u2246e/u2246e02.htm, (accessed 10.02.2014).

Jovanović Popović, M., & Ignjatović, D. (2013a). National Typology of residential buildings in Serbia. Faculty of Architecture, University of Belgrade and GIZ–German Association for International Cooperation, Belgrade. Retrieved from http://www.arh.bg.ac.rs/wp-content/uploads/201415_docs/SAS_EEZA_publikacije/National_Typology_of_residential_buildings_in_Serbia.pdf, (accessed 10.12.2014).

Jovanović Popović, M., & Ignjatović, D. (2013b). Atlas of Multi Family Housing in Serbia. Faculty of Architecture, University of Belgrade and GIZ–German Association for International Cooperation, Belgrade.

Konstantinou, T. (2014). Facade Refurbishment Toolbox: Supporting the Design of Residential Energy Upgrades. Delft University of Technology, Faculty of Architecture and The Build Environment, Architectural Engineering + Technology department.

Krec. (2013). Thermische Gebäudesimulation, retrieved from http://krec.at/index.php?id=21 (accessed 01.12.2013).

Mihailović, Z. (1979). Catalogue of typical house design, Organization for architectural design Naš stan, Belgrade.

Ministry of Energy, Development and Environmental Protection of the Republic of Serbia. (2011). Rulebook on energy efficiency of Buildings. Official Gazette of the Republic of Serbia no. 61/2011. Belgrade.

National Bank of Serbia. (2013). Exchange rates at internet, retrieved from http://www.nbs.rs/export/sites/default/internet/english/scripts/kl_srednji.html, (accessed 03.01.2013).

Neuhoff, K., Amecke, H., Novikova, A., & Stelmakh, K. (2011). Thermal efficiency retrofit of residential buildings: The German experience. CPI Report, Climate Policy Initiative.

OIB. (2011). OIB-Richtlinie 6; Energieeinsparung und Wärmeschutz; Öerreichisches Institut für Bautechnik.

ProCredit Bank. (2013). Energy Efficiency Loans, retrieved from https://www.procreditbank.rs/en/strana/6381/financing-of-energy-efficient-solutions, (accessed 15.04.2013).

PVGIS. (2014). Photovoltaic Geographical Information System, retrieved from http://re.jrc.ec.europa.eu/pvgis/ (accessed 01.12.2014).

Srbija Gas. (2013). Retrieved from http://www.srbijagas.com/, (accessed 15.01.2013).

Statistical Office of Republic of Serbia. (2013). Retrieved from http://www.stat.gov.rs, (accessed 15.01.2013).

Todorović, M. (2010). First NEEAP/BS national energy efficiency action plan/building sector 2009-2018. u: Study Report and NEEAP-BS for the Republic of Serbia Ministry of Mining and Energy. Washington: IRG, June.

Zrnić, S., Ćulum, Z. (1998). Technical Regulations on Heating, Cooling and Air Conditioning. Naučna knjiga, Beograd (in Serbian).

Future Housing Designs

4

4.1 Pursuing Future Designs

The previous chapters reviewed the architecture of traditional housing in Serbia (Chap. 2), followed by examining various scenarios and the associated factors for refurbishment (Chap. 3). These help use to understand the potential of energy efficiency in the Serbian building sector. Knowing this, the next important step is to providing a basis for integrated energy-efficient design in the future.

In this chapter, patterns for the energy-efficient design of prospective buildings in different regions in Serbia will be proposed. The focus will be on climate-appropriate envelopes of residential buildings, keeping in mind the discussion from Chap. 2 on the variations in traditional housing across the country.

The factors that influence the assessment of future housing are first discussed. This is followed by the results of three experimental case studies. First, building thermal and energy-use behavior is examined by a parametric study. Next, the efficiency of different building models for one specific location, namely, Belgrade, is assessed. The third involves the development of models of energy-efficient housing that are suitable for the different regions (and hence, climates) of Serbia and from this the design patterns. Five locations will be considered: Subotica, Belgrade, Užice, Kopaonik Mountain, and Niš.

The following part of this chapter will be concerned with the synergy of energy efficiency measures and renewable energy sources, which is obligatory if one wishes to reach energy-plus levels. Therefore, the potential for the use of photovoltaic solar electricity potential in Serbia will be discussed in the last case study presented in this book.

© Springer Fachmedien Wiesbaden GmbH, part of Springer Nature 2019
V. Jovanović, *Energy-efficient building design in Southeast Europe*,
https://doi.org/10.1007/978-3-658-24165-0_4

4.2   Factors Surrounding the Assessment of Future Housing

As alluded to in the introduction, future buildings will need to have very-high-energy performance in order to meet the targets set by the EU for 2020, 2030, and 2050 (see Table 1.1). The importance of this is apparent when one considers how the building sector is responsible for 40% of the EU's total energy consumption, as well as being a major contributor to GHG emissions. In the case of Serbia, the country's solar energy potential could contribute significantly toward reaching such goals, although at the same time, thought must be given to the optimal thermal comfort conditions that can be sustained throughout the year in an energy-efficient manner.

In this chapter, a series of climate-appropriate building design patterns will be developed for different geographical regions in Serbia, identified by topography, climate, and vegetation. The criteria considered when developing these designs are as follows:

1. Compactness, or the relationship between the building's form and volume. This is the relationship between the building's external surface area to its internal volume or the A/V ratio. This ratio affects the building's energy consumption as the greater the external surface area, the greater the potential for energy loss. For example, buildings made up of materials with the same thermal properties, equivalent levels of air tightness, and the same orientation could have completely different heating demands based solely on their A/V.
2. The standing of the building, which refers to the basic division of houses. This may be classified according to whether it is (1) a free-standing or detached house, (2) a double or semidetached house, or (3) a terraced house or a house in a row. The standing of a house is employed as criterion when improving the energy efficiency of buildings because it has a very strong influence on the early design phase, hence on the building's overall energy efficiency. For example, if a house has a direct neighbor (i.e., houses in a row), then it would have a smaller surface with external walls and therefore lower heat loss to the outside environment.
3. The effect of the orientation of the main façade was examined experimentally, where reference buildings are modeled as being initially oriented toward the south and then rotated toward the east and west. Such experiments allow the differences in the heating demands and the overheating potential of the analyzed buildings to be assessed.
4. The proportion of window glazing on the south-facing façade. The optimal use of window glazing for south-facing façades has long been a prerequisite for energy-efficient houses in continental Europe (BRE 2011). For instance, a passive house requires optimal glazing on south-facing façades and reduced glazing facing north. However, the use of overdesigned glazing may increase the risk of overheating in summer, especially in Southeast European countries. Therefore, glazing is considered an important design parameter that must be evaluated for each case at a time.
5. The type and extent of the insulation of the buildings and the choice of heating, ventilation, and air-conditioning (HVAC) system. These factors are among the most

critical aspects when considering reduced energy consumption and GHG emissions. For example, modifying the thermal properties of the building's various elements from having higher heat-transfer coefficients (U-values) to lower ones (according to the Passivhaus standards) would lead to reduced heating demands in winter.

6. The suitability of different types of construction was evaluated. This referred to building elements with the same heat-transfer coefficients but different thermal storage capacities. The tested types were massive construction and lightweight construction (see below).

7. The prevention of overheating is relevant to the suitability of the buildings to summer conditions. This must take into consideration a building's orientation, type of construction, and proportion of glazing, evaluated in terms of thermal comfort indicators. Passive measures may be employed, such as the installation of internal and external shutters and night ventilation.

Based on the above outlined criteria, a series of three case studies of different means of achieving improved energy performance for Serbia housing will now be presented. The first assessed the seven parameters discussed above to understand their individual importance with regard to improving energy efficiency. The second involved the assessment of three single-family and one multi-family residential buildings in Belgrade under both summer and winter conditions; the third considered a design pattern concept that was suitable for the different regions of Serbia. This will be followed by a case study dealing with the potential contribution of renewable energy resources, in particular solar energy.

4.3   Case Study 7: Energy Performance of Various House Types

The aim of this case study is to investigate the effect various designs of detached family dwellings have on their energy performance. The intention is to gain input that will allow more energy-efficient future designs, which is especially needed at the early design stages, and when considering the depth of understanding required for constructing nZEB (nearly zero-energy buildings) homes.

The dynamic simulations were performed using the Euro-Waebed and GEBA simulation software tools. The reference building was a two-story detached single-family house with a heated floor area of 128 m^2 located in Belgrade (Fig. 4.1). The simulations were carried out for summer and winter conditions, and the analyzed building performance indicators were annual heating demands and the potential for overheating.

The simulations involved changing each of the considered parameters one at a time in order to determine how that specific aspect affected the building's energy behavior. The parameters assessed were the seven outlined in the previous section, namely, compactness, the standing of the house, orientation, proportion of glazing per south-facing façade, the type of insulation level and heating/cooling means, type of construction, and overheating potential. The matrix of the tested options for each parameter is presented in Fig. 4.2.

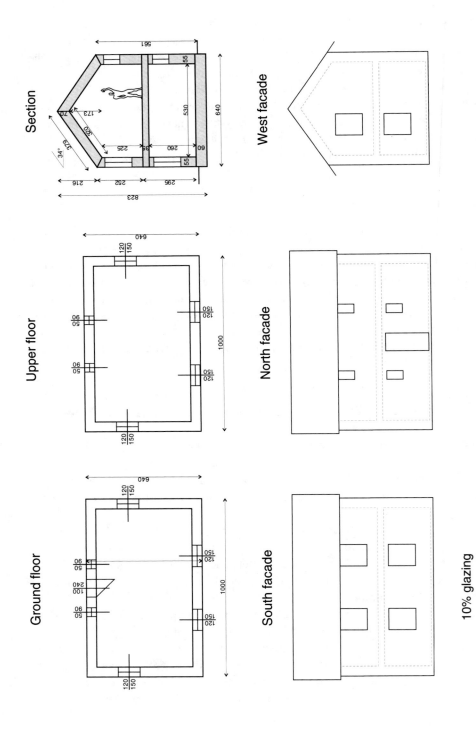

Fig. 4.1 The reference simulation model for the first case study, representing a detached single-family house located in Belgrade, Serbia

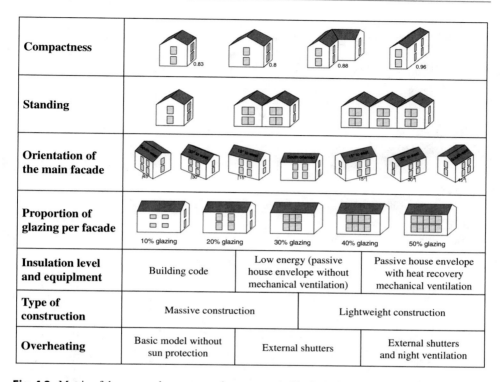

Fig. 4.2 Matrix of the assessed parameters that were varied in the building design and how they were changed (see the text for details)

The compactness was analyzed using four scenarios involving A/V ratios of 0.83, 0.80, 0.88, and 0.96, with the heated floor area kept between 128 and 130 m². The standing of the house was evaluated using the three variants described above, a detached, semidetached, or a row house. The orientations considered involved an initial south-facing orientation, then examples after rotating the house toward the east and west by 15, 30, and 45°. The proportions glazing per the south façade tested were 10%, 20%, 30%, 40%, and 50%. Three insulation options were tested: the building's envelope insulated according to the national building code, involving U-values for the external walls, the roof, and the floor of 0.30 W/m²K; a passive house building envelope with U-values of about 0.11 W/m²K, without energy recovery ventilation; and the same passive house but with 70% efficient energy recovery ventilation. Ventilation of this efficiency was chosen in consideration of the products available on the market and in consultations with local engineers. However, a 95% efficient energy recovery ventilation system was also tested for completeness. The types of construction examined consisted of massive and lightweight construction (Fig. 4.3). The final parameter, overheating prevention, took into consideration various passive cooling methods, i.e., shading, external shutters, and night ventilation.

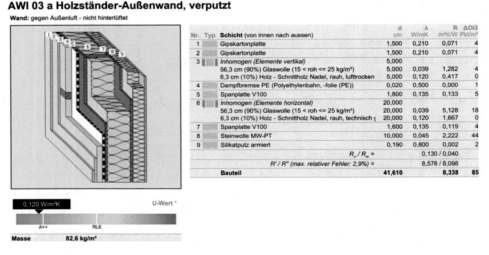

AWm 05 a Hochlochziegel-Außenwand, WDVS

Wand: gegen Außenluft - nicht hinterlüftet

Nr.	Typ	Schicht (von innen nach aussen)	d cm	λ W/mK	R m²K/W	ΔOI3 Pkt/m²
1		Kalk-Zementputz	1,500	1,000	0,015	3
2		Ziegel - Hochlochziegel porosiert <=800kg/m³	25,000	0,250	1,000	35
3		Polystyrol (EPS f. Wärmedämmverbundsysteme WDVS	30,000	0,040	7,500	32
4		Silikatputz armiert	0,190	0,800	0,002	2
		R_{si}/R_{se} =			0,130 / 0,040	
		R'/R'' (max. relativer Fehler: 0,0%) =			8,687 / 8,687	
		Bauteil	**56,690**		**8,687**	**72**

0.115 W/m²K   U-Wert [1]

A++   RL6

Masse   235,8 kg/m²

AWl 03 a Holzständer-Außenwand, verputzt

Wand: gegen Außenluft - nicht hinterlüftet

Nr.	Typ	Schicht (von innen nach aussen)	d cm	λ W/mK	R m²K/W	ΔOI3 Pkt/m²
1		Gipskartonplatte	1,500	0,210	0,071	4
2		Gipskartonplatte	1,500	0,210	0,071	4
3		Inhomogen (Elemente vertikal)	5,000			
		56,3 cm (90%) Glaswolle (15 < roh <= 25 kg/m³)	5,000	0,039	1,282	4
		6,3 cm (10%) Holz - Schnittholz Nadel, rauh, lufttrocken	5,000	0,120	0,417	0
4		Dampfbremse PE (Polyethylenbahn, -folie (PE))	0,020	0,500	0,000	1
5		Spanplatte V100	1,800	0,135	0,133	5
6		Inhomogen (Elemente horizontal)	20,000			
		56,3 cm (90%) Glaswolle (15 < roh <= 25 kg/m³)	20,000	0,039	5,128	18
		6,3 cm (10%) Holz - Schnittholz Nadel, rauh, technisch	20,000	0,120	1,667	0
7		Spanplatte V100	1,600	0,135	0,119	4
8		Steinwolle MW-PT	10,000	0,045	2,222	44
9		Silikatputz armiert	0,190	0,800	0,002	2
		R_{si}/R_{se} =			0,130 / 0,040	
		R'/R'' (max. relativer Fehler: 2,9%) =			8,578 / 8,098	
		Bauteil	**41,610**		**8,338**	**85**

0.120 W/m²K   U-Wert [1]

A++   RL6

Masse   82,6 kg/m²

Fig. 4.3 Top: Massive type of external wall construction. Bottom: Lightweight type of the external wall construction (Screenshots from online-platform, Source: IBO 2013)

Considering first the effect of varying the compactness of the building or the A/V ratio, it was found that the most compact house (meaning the smallest external surface area to interior volume) had the lowest heating demands, as depicted in Fig. 4.4. It can be seen that the difference in heating demands between the most and least compact models was 6 kWh/m²a, or around 10%.

Next, the simulation of the standing of the house indicated that the option of a house in a row had the best energy efficiency in terms of house-heating demands (Fig. 4.5), being 17 kWh/m²a more efficient than the detached house, which showed the worse performance.

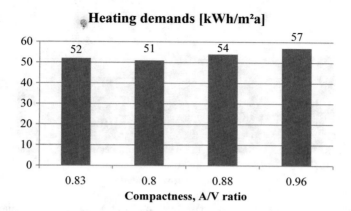

Fig. 4.4 Heating demands of four similar houses with different compactness values (A/V ratios)

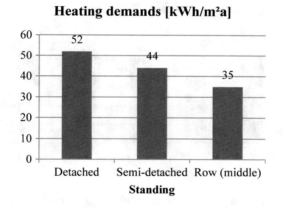

Fig. 4.5 Heating demands of three similar houses with different standing

Increasing the glazing per main (south-facing) façade showed a gradual decrease in the heating demands, although with diminishing returns at the higher percentages. We see an increase from 10 to 20% in glazing led to a reduction in demand of 5 kWh/m^{2}a (Fig. 4.6), although this decreases to 2 kWh/m^{2}a for the increase from 40 to 50%. The total decrease in heating demand after increasing the glazing from 10 to 50% was 14.5 kWh/m^{2}a.

The orientation of the main façade appears to have a negligible effect on the energy efficiency of the house. We see that the difference in the heating demands between the minimum (a south-facing façade) and the maximum, where the façade is orientated at 45° (south-west and south-east; see Fig. 4.7), was only 3 kWh/m^{2}a.

Consideration of the type or level of insulation showed the greatest potential for saving energy, as seen in the previous chapter. It is found (Fig. 4.8) that the passive house level of insulation without energy recovery ventilation results in a reduction of heating demands by

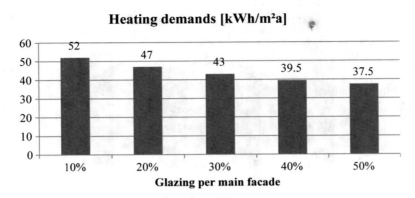

Fig. 4.6 Heating demands of five similar houses with different percentages of glazing per south-facing façade

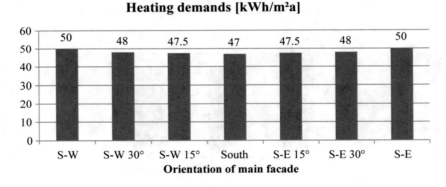

Fig. 4.7 Heating demands of seven similar houses with different orientations of the main façade

32 kWh/m²a, from 52 kWh/m²a to 20 kWh/m²a, while including the ventilation results in the "factor of 10" improvement, from 52 kWh/m²a to 5 kWh/m²a.

The analysis of the effect of the two types of construction indicated that massive construction performed better in terms of summer suitability, which in this case means lower temperatures in rooms with a potential for overheating (Fig. 4.9). By contrast, the lightweight construction showed a greater tendency for overheating, even with minimal glazing on the south façade. This carries over to the final assessment, dealing with the prevention of overheating, also presented in Fig. 4.9. Such concerns are critical for Belgrade, given its tendency for hot summers (as well as for other locations in Serbia). For the basic house, the internal air temperatures reached 32.4 °C for the massive construction and 33 °C for the lightweight construction, even when only using minimal

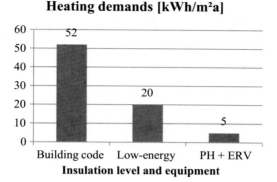

Heating demands [kWh/m²a]

Fig. 4.8 Heating demands of three similar houses with different insulation levels and HVAC equipment (*PH* passive house, *ERV* energy recovery ventilation)

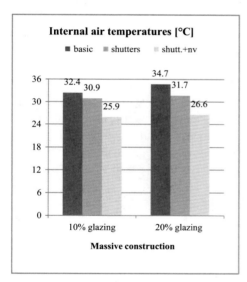

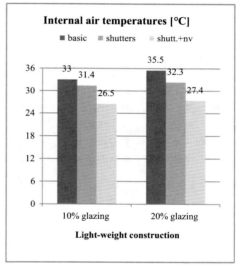

Fig. 4.9 Internal air temperatures in similar rooms with a potential for overheating for the massive and lightweight construction types. Two modifications for south-glazing were tested, 10 and 20% glazing, as well as the use of shutters, plus shutters, and night-time ventilation

glazing. Adding external shutters reduced these temperatures by an average of 2 °C, while only including both shutters and night ventilation allows the temperatures to fall below an acceptable 27 °C.[1]

---

[1]Note that thermal comfort is usually set to 20–26°, although in Austria, this is increased to 27° for summer months, which is considered to still be acceptable for a passive house.

These results demonstrated that each of the assessed parameters has some influence on the building's energy efficiency, although with differing degrees. For example, the type of insulation used made the greatest different, as alluded to in the refurbishment chapter, while the orientation of the house was relatively minor. In the following case study, four different basic types of residential buildings in Belgrade will be examined, again using the seven parameters just assessed.

4.4   Case Study 8: Comparison of Four Types of Residential Buildings

The aim of this case study is to evaluate the energy consumption performance of four types of residential buildings in Belgrade, Serbia. The buildings were assessed under winter and summer regimes, and the measures used to evaluate their performances were heating demands and internal air temperatures, respectively.

Two simulation software packages were used, EW and GEBA. The housing examples considered were three different types of single-family house and one type of multistory residential building. The first house (SR, single house with a pitched roof) was a two-story building with a heated zone of area 128 m^2 (Fig. 4.10). The second house (SF, single house with a flat roof) was a two-story building, again with a heated area of 128 m^2 (Fig. 4.11). The third example (SB, single house with a pitched roof and basement) was a two-story building with a heated area of 160 m^2 and a basement and attic, both of which were non-heated zones (Fig. 4.12). The multi-family building (MH, multi-family with a flat roof) was a six-story structure (Fig. 4.13) with a heated area of 1895 m^2.

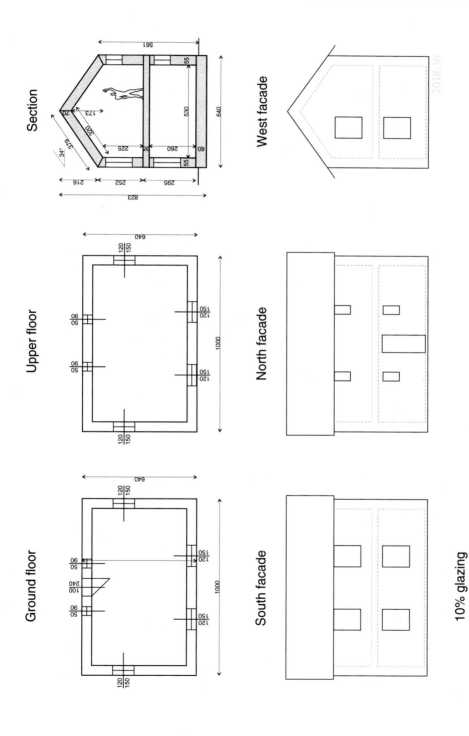

Fig. 4.10 Model SR: A single-family house of 128 m² heated floor area, with a pitched roof

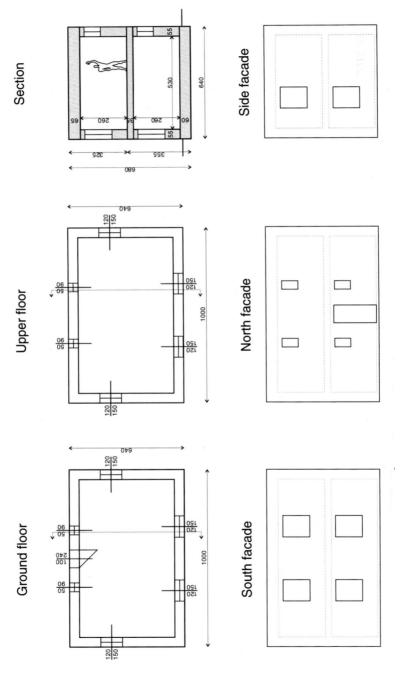

Fig. 4.11 Model SF: Single house of 128 m^2 heated floor area, with a flat roof

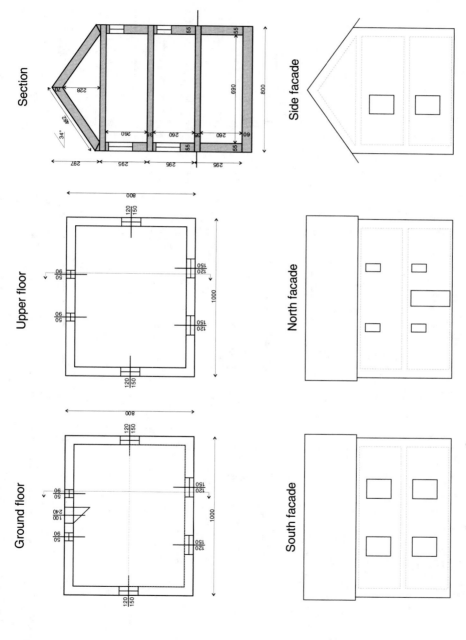

Fig. 4.12 Model SB: Single house of 160 m² heated floor area with a pitched roof and a basement and attic as non-heated area

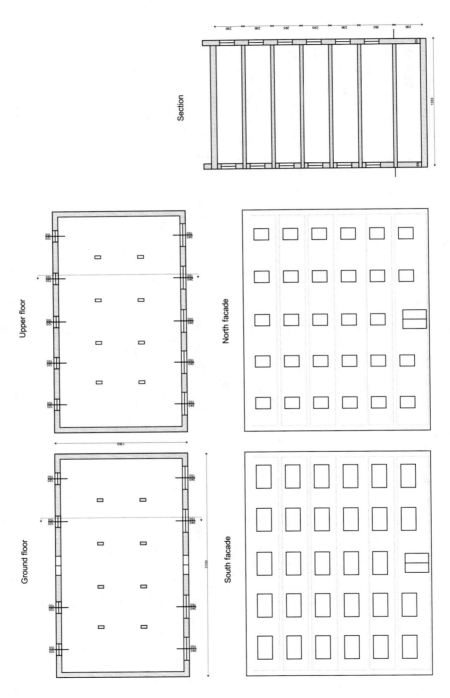

Fig. 4.13   Model MH: Multistory (six floors) multi-family housing of 1895 m² heated floor area with a flat roof

The buildings were first modeled using the basic setting; then as in the previous section, the first three houses were modeled while adjusting the seven parameters presented above (see Fig. 4.4). However, for the multi-family dwelling, fewer options were explored. Here, standing was set to row, only the south-facing façade option was used, the glazing options were restricted to 10 and 20%, only massive construction was considered, and for the potential overheating, the options were no shutters, shutters, and shutters plus night ventilation (NV). The outcomes of these studies will then be able to be exploited as input data for further architectural design developments.

Starting the discussion from the house SR (Fig. 4.10), the winter mode evaluation identified a number of differences in the heating demand resulting from the different design options (Table 4.1). As found before, the orientation of the house appeared to have little effect under winter conditions, with the class of insulation and HVAC system showing the greatest effect. While the higher levels of glazing per south façade appears to reduce heating demands, it is not considered a good option owing to the fact that the higher glazing levels also lead to increased possibility of overheating in summer, where even small levels of glazing appear to lead to a high overheating potential (Table 4.2). This means that the decreased heating demands in winter will be somewhat offset by the increased cooling needs in summer. Furthermore, when considering summer conditions, the massive type of construction appears to show better results than the lightweight construction, with lower room temperatures.

Table 4.1 Heating demands of the SR model (Fig. 4.10) resulting from different design parameters (Fig. 4.2) considering winter operating conditions

Design parameter	Settings	Heating demands [kWh/m²a]
Compactness, A/V ratio	0.83	52.25
	0.8	51.00
	0.88	54.26
	0.96	57.4
Standing	Detached house	52.25
	Semidetached house	43.66
	Row house (middle)	35.26
Orientation of main façade	South-west orientation	49.66
	30° toward west	48.41
	15° toward west	47.48
	South-facing orientation	47.17
	15° toward east	47.48
	30° toward east	48.41
	South-east orientation	49.67
Glazing per south façade	10%	52.25
	20%	47.17
	30%	42.94
	40%	39.43
	50%	37.52
Insulation level and HVAC	Building code	52.25
	Low energy house (PH envelope)	19.72
	PH envelope + 70% ERV	5.18
	PH envelope + 90% ERV	2.09

Shaded fields indicate the options with good performances
PH passive house, *ERV* energy recovery ventilator

Table 4.2 Internal air temperatures in potentially overheated rooms of the SR model (Fig. 4.10) considering the different design parameters (Fig. 4.2)

Type of construction	Glazing per south façade (%)	Overheating prevention	Max. temperature in rooms (°C)
Massive construction	10	No sun protection	32.4
		External shutters	30.9
		Shutters + NV	25.9
	20	No sun protection	34.7
		External shutters	31.7
		Shutters + NV	26.6
	30 and more	Shutters + NV	Above 27
Lightweight construction	10	No sun protection	33.0
		External shutters	31.4
		Shutters + NV	26.5
	20	No sun protection	35.5
		External shutters	32.3
		Shutters + NV	27.4
	30 and more	Shutters + NV	Above 28

Shaded fields indicate options with acceptable internal air temperatures
NV night ventilation

Considering now the model SF performs under winter conditions, similar results in the heating demands' dependence on the tested parameters were obtained as with the previous model (Table 4.3). Furthermore, the summer mode results for this house model revealed the overheating potential of the window glazing, as well as again showing the massive construction providing better results in terms of overheating prevention (Table 4.4).

Table 4.3 Heating demands of the SF model (Fig. 4.11) resulting from the different design parameters (Fig. 4.2) under winter conditions

Design parameter	Settings	Heating demands [kWh/m²a]
Compactness, A/V ratio	0.83	51.45
	0.8	50.55
	0.88	54.82
	0.96	58.80
Standing	Detached house	51.45
	Semidetached house	41.60
	Row house (middle)	32.27
Orientation of main façade	South-west oriented	49.39
	30° toward west	48.10
	15° toward west	47.21
	South oriented	46.9
	15° toward east	47.22
	30° toward east	48.10
	South-east oriented	49.36
Glazing per south façade	10%	51.45
	20%	46.90
	30%	42.16
	40%	38.68
	50%	37.03
Insulation level and HVAC	Building code	51.45
	Low energy house (PH envelope)	18.82
	PH envelope + 70% ERV	5.02
	PH envelope + 90% ERV	2.08

Shaded fields indicate the options with good performances
PH passive house, *ERV* energy recovery ventilator

Table 4.4 Internal air temperatures in potentially overheated rooms of the SF model (Fig. 4.11) considering the different design parameters (Fig. 4.2)

Type of construction	Glazing per south façade (%)	Overheating prevention	Max. temperature in rooms (°C)
Massive construction	10	No sun protection	33.1
		External shutters	31.3
		Shutters + NV	26.0
	20	No sun protection	35.7
		External shutters	32.1
		Shutters + NV	26.8
	30 and more	Shutters + NV	Above 27
Lightweight construction	10	No sun protection	33.9
		External shutters	32.0
		Shutters + NV	26.9
	20	No sun protection	36.8
		External shutters	33.0
		Shutters + NV	28.1
	30 and more	Shutters + NV	Above 28

Shaded fields indicate options with acceptable internal air temperatures
NV night ventilation

For model SB, the winter mode assessment results are listed in Table 4.5, while the summer time results are in Table 4.6. The summer mode pointed out overheating potential of both massive and lightweight construction.

Table 4.5 Heating demands of the SB model (Fig. 4.12) resulting from the different design parameters (Fig. 4.2) under winter conditions

Design parameter	Settings	Heating demands [kWh/m^{2}a]
Compactness, A/V ratio	0.83	52.58
Standing	Detached house	52.58
	Semidetached house	45.21
	Row house (middle)	38.53
Orientation of main façade	South-west oriented	50.18
	30° toward west	49.20
	15° toward west	48.47
	South oriented	42.35
	15° toward east	48.47
	30° toward east	49.19
	South-east oriented	50.19
Glazing per south façade	10%	52.58
	20%	48.19
	30%	44.48
	40%	41.25
	50%	38.50
Insulation level and HVAC	Building code	52.58
	Low energy house (PH envelope)	19.12
	PH envelope + 70% ERV	5.62
	PH envelope + 90% ERV	2.66

Shaded fields indicate the options with good performances
PH passive house, *ERV* energy recovery ventilator

Table 4.6 Internal air temperatures in potentially overheated rooms of the SB model (Fig. 4.12) considering different design parameters (Fig. 4.2)

Type of construction	Glazing per south façade (%)	Overheating prevention	Max. temperature in the rooms (°C)
Massive construction	10	No sun protection	33.1
		External shutters	30.5
		Shutters + NV	25.7
	20	No sun protection	35.3
		External shutters	31.3
		Shutters + NV	26.4
	30 and more	Shutters + NV	Above 27
Lightweight construction	10	No sun protection	33.7
		External shutters	30.8
		Shutters + NV	26.3
	20	No sun protection	36.1
		External shutters	31.7
		Shutters + NV	27.2
	30 and more	Shutters + NV	Above 28

Shaded fields indicate options with acceptable internal air temperatures
NV night ventilation

The final house assessed in this section, the multistory house model MH, which can house multiple families, was analyzed using a reduced number of parameters (see above and Table 4.7). As with the previous examples, there appears to be significant overheating potential when using the massive construction option (Table 4.8).

Table 4.7 Heating demands of MH model in Belgrade, under different design parameters

Design parameter	Settings	Heating demands [kWh/m²a]
Standing	Row building (middle)	14.95
Glazing per south façade	10%	14.95
	20%	13.49
Insulation level and HVAC	Low energy house (PH envelope)	14.95

PH passive house

Table 4.8 Internal air temperatures in potentially overheated rooms of the MH model in Belgrade, under different design parameters

Type of construction	Glazing per south façade (%)	Overheating prevention	Max. temperature in the rooms (°C)
Massive construction	10	No sun protection	31.4
		External shutters	30.2
		Shutters + NV	25.5
	20	No sun protection	33.5
		External shutters	30.8
		Shutters + NV	26.0
	30 and more	Shutters + NV	Above 27

Shaded fields indicate options with acceptable internal air temperatures
NV night ventilation

It would be obvious to the reader that the general forms of the results presented above are very similar. This confirms that the employed models were not "extreme cases" but behave more or less in a similar manner for those particular climatic conditions. Following from this, the next case study will examine new design patters that are adapted to the different regions of Serbia.

4.5   Case Study 9: New Design Patterns for Prospective Housing in Different Regions of Serbia

The purpose of this case study is to deliver design patterns for prospective houses adapted to the different regions of Serbia. The patterns are developed in a conceptual form to serve as a database for proposed housing designs and therefore will serve as a pre-step when designing energy-efficient housing in Serbia (and by extension, the wider Southeastern European region).

The dynamic simulations were performed using the Euro-Waebed and GEBA software tools. The standard house for the simulations is the above-described SR single-family house (Fig. 4.10). Five locations were investigated (Fig. 4.14): (1) Subotica, Northern Serbia; (2) Belgrade, the capital; (3) Užice (Zlatibor District), Western Serbia; (4) Kopaonik Mountain, Central Serbia; and (5) Niš, Southeast Serbia. These locations are representative of the varied geography of the country while also reflecting the different types of traditional

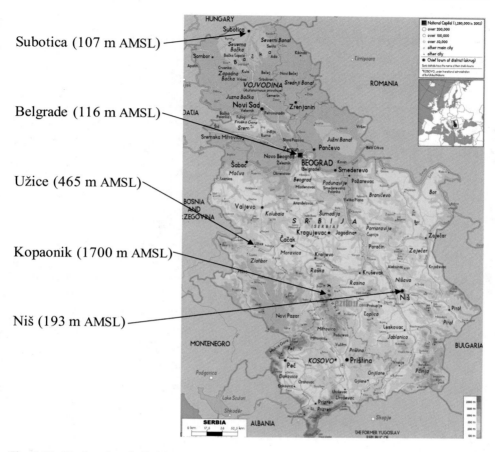

Subotica (107 m AMSL)

Belgrade (116 m AMSL)

Užice (465 m AMSL)

Kopaonik (1700 m AMSL)

Niš (193 m AMSL)

Fig. 4.14 The locations in Serbia (and their heights above mean sea level) that serve as the test sites for the new design patterns

architecture found in Serbia as described in Chap. 2. Aspects of the local climates of each of these locations (namely, their mean maximum and minimum monthly temperatures throughout the year) are presented in Figs. 4.15, 4.16, 4.17, 4.18 and 4.19.

The simulations followed a similar scheme to that presented in the previous test cases. These criteria were again arranged in the form of a design matrix, Table 4.9 outlining the tested design options. Simulations were again done for winter and summer conditions, where the winter mode set out to estimate the heating demands of the house, while the summer mode investigated the suitability of the considered construction type in terms of overheating potential.

The proposed design patterns divided the considered options into recommended, acceptable, and inappropriate for the proposed energy-efficient housing (or nZEBs) for each location (Fig. 4.14). Table 4.10 displays an example of a recommended, an acceptable, and an inappropriate design pattern.

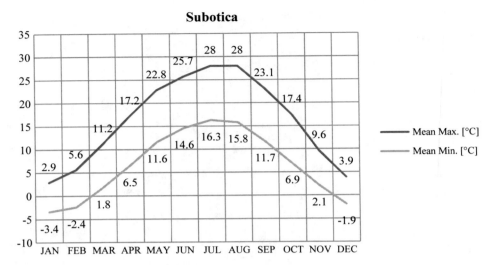

Fig. 4.15 Mean maximal and minimal air temperatures in Subotica for the period 1981–2010.
Source: Republic Hydrometeorological Service of Serbia (2013)

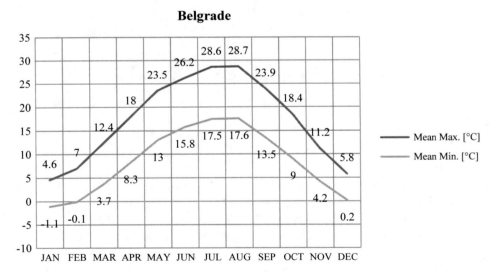

Fig. 4.16 Mean maximal and minimal air temperatures in Belgrade for the period 1981–2010.
Source: Republic Hydrometeorological Service of Serbia (2013)

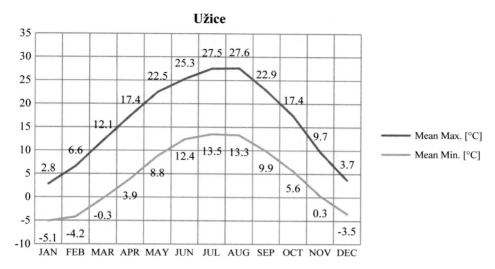

Fig. 4.17 Mean maximal and minimal air temperatures in Užice for the period 1981–2010. The shown mean temperatures are from the meteorological station Požega, located around 16 km from Užice. In the simulations, PVGIS climate data for Užice were used. Source: Republic Hydrometeorological Service of Serbia (2013)

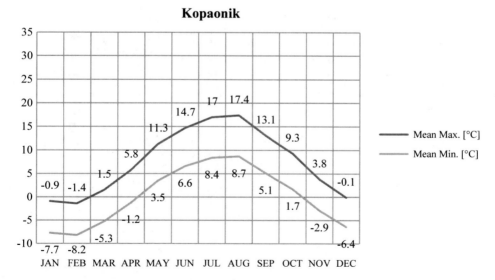

Fig. 4.18 Mean maximal and minimal air temperatures in Kopaonik for the period 1981–2010. Source: Republic Hydrometeorological Service of Serbia (2013)

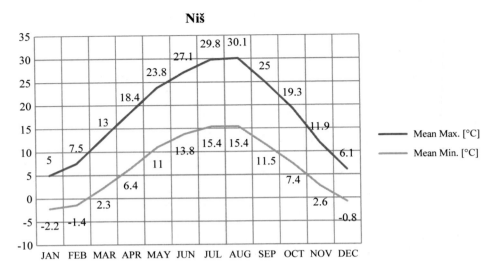

Fig. 4.19 Mean maximal and minimal air temperatures in Niš for the period 1981–2010. Source: Republic Hydrometeorological Service of Serbia (2013)

Table 4.9 The design pattern concepts evaluated through different options of the seven parameters employed throughout this chapter

Compactness	Standing	Orientation main façade	Glazing - south façade	Insulation and equipment	Type of construction	Overheating prevention
Low compact	Detached	South	10%	Building code	Light-weight	No protection
Medium compact	Semi-detached	Southeast	20%	LE (PH, no ERV)	Massive	Ex. shutters
Very compact	Row	Southwest	30%	PH + ERV		Shutters + NV

The obtained results for the winter mode (Table 4.11) show that the southern orientation was a good option for each location. In Subotica and Užice, increased glazing had the smallest impact on heating demands; in Belgrade and Niš, an increase in the glazing percentage led to a heating demand reduction of up to 32%, while increased glazing saw the best results (>50% reduction) occurring for Kopaonik.

Considering now the summer mode, we see that the massive type of construction showed good results when considering the internal temperatures of a potentially overheated

Table 4.10 Examples of recommended, acceptable, and inappropriate design patterns for a nearly zero-energy house

Compactness	Standing	Orientation main façade	Glazing - south façade	Insulation and equipment	Type of construction	Overheating prevention
Low compact	Detached	South	10%	Building code	Light-weight	No protection
Medium compact	Semi-detached	Southeast	20%	LE (PH, no ERV)	Massive	Ex. shutters
Very compact	Row	Southwest	30%	PH + ERV		Shutters + NV

Shaded fields mean "recommended options," blank fields mean "acceptable options," and strike-through fields mean "unfavorable options"
PH passive house, ERV energy recovery ventilation, NH night ventilation

room. In Subotica and Belgrade, 10 to 20% of glazing enabled comfortable conditions in such rooms. In Užice and Kopaonik, each of the tested options provided desirable levels of thermal comfort. However, in Niš, only the minimal glazing resulted in internal air temperatures below 27 °C (Table 4.12).

Considering now the lightweight construction option (Table 4.13), again for the summer conditions, a greater potential for overheating was observed than for the massive construction (Table 4.12). In Subotica and Belgrade, only the south-facing minimally glazed house performed well, while in Užice, options with 10–20% of south-facing glazing resulted in suitable conditions. Only in Kopaonik did the lightweight construction with high levels of south-facing glazing show good results. By contrast, in Niš, all simulated options resulted in overheating.

Based on these simulations, the extensive amount of output data could be presented in such a way as to provide guidelines for the development of climate-appropriate designs for the different regions of Serbia. Hence, Tables 4.14, 4.15, 4.16, 4.17 and 4.18 present for the five considered locations the recommended, acceptable, and inappropriate design options for energy-efficient housing (or nZEBs).

A recommended design pattern for energy-efficient houses in Subotica (Table 4.14) and Belgrade (Table 4.15) was a compact house in a row (note: this was in fact found for all of the test locations), with a south-orientated façade (the same for each location) with up to 20% glazing. The house is recommended to be built in a massive construction style with super insulation (when $U = 0.10$ W/m^{2}K), energy recovery ventilation, and passive cooling measures, such as external shutters and night ventilation.

Table 4.11 Heating demands for the SR model [kWh/m²a] (Fig. 4.10) under winter conditions

Location	Subotica			Belgrade			Užice			Kopaonik			Niš		
Glazing	v1	v2	v3	v1	v2	v3	v1	v2	v3	v1	v2	v3	v1	v2	v3
SW-135°	6.93	6.95	7.14	5.66	5.44	5.38	6.84	6.78	6.88	13.28	6.78	6.88	5.32	5.06	4.96
150°	6.70	6.38	6.29	5.41	4.80	4.43	6.61	6.19	5.99	11.81	6.19	5.99	5.08	4.45	4.05
165°	6.55	6.00	5.73	5.24	4.38	3.81	6.44	5.79	5.40	10.80	5.79	5.40	4.91	4.04	3.47
S-180°	6.50	5.86	5.56	5.17	4.23	3.61	6.39	5.66	5.19	10.44	5.66	5.19	4.85	3.90	3.29
195°	6.55	5.99	5.72	5.24	4.38	3.81	6.44	5.78	5.39	10.78	5.78	5.39	4.91	4.03	3.47
210°	6.70	6.37	6.27	5.41	4.79	4.41	6.60	6.18	5.99	11.80	6.18	5.99	5.08	4.44	4.03
SE-225°	6.93	6.93	7.11	5.65	5.42	5.35	6.83	6.77	6.87	6.83	6.77	6.87	5.32	5.04	4.94
Comment	→ up to 15% decrease			→ up to 30% decrease			→ up to 19% decrease			→ up to 50% decrease			→ up to 32% decrease		

The considered parameters were glazing per façade (v1, 10%; v2, 20%; v3, 30%) and orientation of the main façade (ranging from the southeast toward southwest in 15° steps). The comment row points out the percentage up to which the heating demand is reduced for each location

Table 4.12 Internal air temperatures in potentially overheated rooms considering the massive construction option for the SR model [°C] (Fig. 4.10) under summer conditions

Location	Subotica			Belgrade			Užice			Kopaonik			Niš		
Glazing	v1	v2	v3	v1	v2	v3	v1	v2	v3	v1	v2	v3	v1	v2	v3
SW-135°	26.0	26.8	27.8	26.0	26.8	27.7	25.0	25.8	26.7	22.0	24.7	27.4	26.9	27.7	x
150°	25.9	26.6	27.6	25.9	26.7	27.7	25.0	25.7	26.7	21.6	24.0	26.3	26.9	x	x
165°	25.9	26.5	27.4	25.9	26.6	27.5	24.9	25.6	26.5	21.3	23.5	25.6	26.8	x	x
S-180°	25.9	26.6	27.5	25.9	26.6	27.4	24.9	25.6	26.4	21.2	23.4	25.5	26.8	27.5	28.4
195°	25.9	26.6	27.5	25.9	26.6	27.5	24.9	25.6	26.5	21.5	23.9	26.2	26.8	x	x
210°	26.0	26.7	27.7	26.0	26.7	27.7	25.0	25.7	26.7	22.0	24.8	27.6	26.9	x	x
SE-225°	26.0	26.9	28.0	26.0	26.9	28.0	25.1	25.9	27.0	22.5	25.9	29.1	27	27.7	x

The considered parameters were glazing per façade (v1, 10%; v2, 20%; v3, 30%) and the orientation of the main façade (ranging from the southeast toward southwest in 15° steps). The shading indicates acceptable conditions

Table 4.13 Internal air temperatures in potentially overheated rooms considering the lightweight construction option for the SR model [°C] (Fig. 4.10) under summer conditions

Location	Subotica			Belgrade			Užice			Kopaonik			Niš		
Glazing	v1	v2	v3	v1	v2	v3	v1	v2	v3	v1	v2	v3	v1	v2	v3
SW-135°	26.7	x	x	26.7	x	x	25.7	26.7	x	22.5	25.4	28.6	27.6	x	x
150°	26.6	x	x	26.6	x	x	25.7	26.7	x	22.2	24.7	27.4	27.6	x	x
165°	26.6	x	x	26.6	x	x	25.6	26.5	x	21.9	24.2	26.6	27.5	x	x
S-180°	26.6	27.5	28.6	26.5	27.4	28.5	25.6	26.4	27.5	21.8	24.2	26.6	27.5	28.5	29.5
195°	26.6	x	x	26.6	x	x	25.6	26.5	x	22.1	24.8	27.5	27.5	x	x
210°	26.6	x	x	26.6	x	x	25.7	26.6	x	22.8	26.0	29.1	27.6	x	x
SE-225°	26.7	x	x	26.7	x	x	25.8	26.9	x	23.4	27.3	30.9	27.7	x	x

The considered parameters were glazing per façade (v1, 10%; v2, 20%; v3, 30%) and the orientation of the main façade (ranging from the southeast toward southwest in 15° steps)

Table 4.14 Design patterns for energy-efficient houses in Subotica

Compactness	Standing	Orientation main façade	Glazing - south façade	Insulation and equipment	Type of construction	Overheating prevention
~~Less compact~~	Detached	South	10%	~~Building code~~	~~Light-weight~~	~~No protection~~
Medium compact	Semi-detached	Southeast	20%	LE (PH, no ERV)	Massive	Ex. shutters
Very compact	Row	Southwest	~~30%~~	PH + ERV		Shutters + NV

Shaded, recommended; blank, acceptable; strike-through, unfavorable

Table 4.15 Design patterns for energy-efficient houses in Belgrade

Compactness	Standing	Orientation main façade	Glazing - south façade	Insulation and equipment	Type of construction	Overheating prevention
~~Less compact~~	Detached	South	10%	~~Building code~~	~~Light-weight~~	~~No protection~~
Medium compact	Semi-detached	Southeast	20%	LE (PH, no ERV)	Massive	Ex. shutters
Very compact	Row	Southwest	~~30%~~	PH + ERV		Shutters + NV

Shaded, recommended; blank, acceptable; strike-through, unfavorable

For Užice (Table 4.16), many of the suggestions for the previous houses are the same, although 30% glazing and lightweight construction were also acceptable.

In the case of Kopaonik (Table 4.17), more than 10% of glazing per south façade was considered acceptable, with either massive or lightweight construction, super insulation, and energy recovery ventilation, although without passive cooling measures.

Finally for Niš, the resulting pattern (Table 4.18) was quite similar to that in Subotica and Belgrade, except for the use of reduced glazing of the southern façade, owing to here showing the greatest potential for overheating.

Table 4.16 Design patterns for energy-efficient houses in Užice

Compactness	Standing	Orientation main façade	Glazing - south façade	Insulation and equipment	Type of construction	Overheating prevention
~~Less compact~~	Detached	South	10%	~~Building code~~	Light-weight	~~No protection~~
Medium compact	Semi-detached	Southeast	20%	LE (PH, no ERV)	Massive	Ex. shutters
Very compact	Row	Southwest	30%	PH + ERV		Shutters + NV

Shaded, recommended; blank, acceptable; strike-through, unfavorable

Table 4.17 Design patterns for energy-efficient houses on Kopaonik

Compactness	Standing	Orientation main façade	Glazing - south façade	Insulation and equipment	Type of construction	Overheating prevention
~~Less compact~~	Detached	South	~~10%~~	~~Building code~~	Light-weight	No protection
Medium compact	Semi-detached	Southeast	20%	LE (PH, no ERV)	Massive	Ex. shutters
Very compact	Row	Southwest	30%	PH + ERV		~~Shutters + NV~~

Shaded, recommended; blank, acceptable; strike-through, unfavorable

Table 4.18 Design patterns for energy-efficient houses in Niš (shaded, recommended; blank, acceptable; strike-through, unfavorable)

Compactness	Standing	Orientation main façade	Glazing - south façade	Insulation and equipment	Type of construction	Overheating prevention
Less compact	Detached	South	10%	Building code	Light-weight	No protection
Medium compact	Semi-detached	Southeast	20%	LE (PH, no ERV)	Massive	Ex. shutters
Very compact	Row	Southwest	30%	PH + ERV		Shutters + NV

4.6   Case Study 10: Renewable Energy Sources: Solar Energy Potential

As demonstrated by the three case studies presented so far in this chapter, it is possible to build a highly energy-efficient house in Serbia. This requires the selection of the optimal combinations of the seven assessed criteria (although these are not the only ones possible) and considering the specific environment where the dwelling will be build. However, to reach nZEB levels of energy efficiency, renewable energy sources have to be utilized. Renewable energy is the energy that can be produced from sources that are renewed through ongoing, repetitive natural processes, such as solar radiation, wind, water flow, geothermal, and biomass. Both heat and electric energy can be produced from renewable energy sources; therefore, the synergy of energy conservation measures and the development of renewable energy sources are essential for buildings to reach energy plus levels.

Fortunately, the renewable energy potential in Serbia is great, in particular solar energy. As outlined in Chap. 1, solar radiation levels are about 40% higher in Serbia than the European average (see Fig. 4.20) and about 30% higher than in Central Europe (Stamenić 2010). However, the potential of this resource has not yet been exploited in Serbia (up until 2014).

In the following, the potential for solar energy being used in Serbian households will be explored. The efficiency of photovoltaic (PV) technology or solar cells is largely dependent upon two factors: the type of solar cells used and the orientation of the PV modules.

Considering first the types of PV cells, these can be divided into two basic kinds: crystalline modules (monocrystalline and polycrystalline) and thin-film modules. Monocrystalline PV cells are made from thin slices of a single crystal, which makes them the

Global irradiation and solar electricity potential
Optimally-inclined photovoltaic modules

SERBIA / СРБИЈА

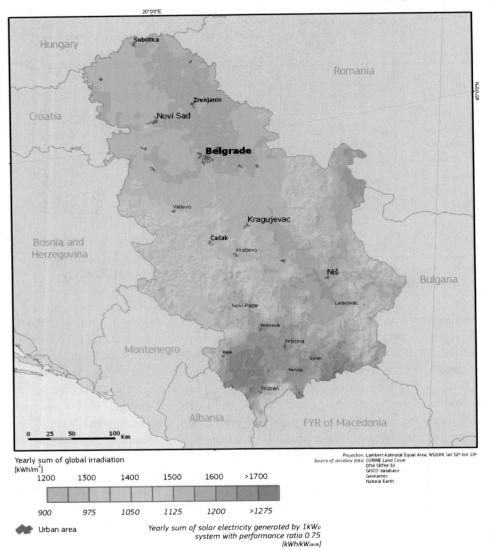

Fig. 4.20 Global irradiation and solar electricity potential for optimally-inclined photovoltaic modules in Serbia (Adopted from PVGIS)

most efficient, although also the most expensive. The efficiency of monocrystalline PV panels varies from 15 to 20%. Polycrystalline PV are made from multi-crystalline material, leading to their production being less expensive, although they are also less efficient than monocrystalline cells, with values of between 13 and 16%. Thin-film solar panels are produced from extra thin layers of a semiconducting material. While these panels are less expensive that the other types just discussed, they are again less efficient, with values ranging from 6 to 12%. Such thin-film cells could be used to create building materials, such as building-integrated photovoltaics (BIPV). The main types of thin-film cells are cadmium telluride cells (CdTe), copper indium gallium diselenide (CIGS), amorphous silicon (a-Si), and thin-film silicon (TF-Si). Table 4.19 compares the efficiency of some of these different types of PV cells.

Concerning the optimal orientation of PV modules in Serbia, this is a function of a location's latitude. According to the PVGIS database, for Serbia, this is toward the south, with an angle ranging from 30° to 35° from the horizontal (see Table 4.20).

In the following case study, the aim is to examine the potential of PV energy generation in the five locations introduced above (Fig. 4.14). The focus is on grid-connected crystalline silicon PV technology with a peak power when installed of 1 kWp. The potential is modeled using the PVGIS simulation software (see Fig. 4.21 for a screenshot of this tool for the case of Belgrade). For each location, two options for mounting the PV cells are tested: a free-standing PV system and BIPV. Furthermore, optimal angles to gain the maximum efficiency from the cells are determined.

The calculations pointed out that, naturally, south-oriented PV modules were required throughout the country. PV modules inclined from 33° to 35° performed well in the

Table 4.19 Approximate efficiency of different PV cells

Cell type	Monocrystalline (%)	Polycrystalline (%)	CdTe (%)	CIGS (%)	a-Si (%)
Module efficiency	15–20	13–16	9–11	10–12	6–8

Table 4.20 Annual electricity production and optimal mounting options for grid-connected crystalline silicon PV systems of a nominal power of 1 kW

Location	Subotica	Belgrade	Užice	Kopaonik	Niš
Optimal orientation	South	South	South	South	South
Free-standing: Inclination	35°	34°	32°	32°	33°
BIPV: Inclination	34°	33°	31°	31°	32°
Free-standing: Output	1170 kWh/a	1180 kWh/a	1100 kWh/a	1170 kWh/a	1200 kWh/a
BIPV: Output	1100 kWh/a	1120 kWh/a	1040 kWh/a	1110 kWh/a	1130 kWh/a

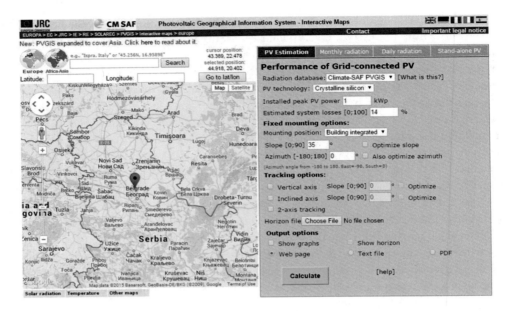

Fig. 4.21 A screenshot from the PVGIS simulation tool for building-integrated photovoltaics (BIPV) in Belgrade (Screenshot from PVGIS online-platform, Source: European Commission)

northern part of the country (e.g., Subotica, Belgrade), while in the south (e.g., Užice, Kopaonik, Niš), the optimal inclination was from 31° to 33°. The free-standing PV system appears to have performed slightly better than the BIPV. PV electricity potential was very similar for each of the analyzed locations, with the highest potential in in Niš southeast Serbia, although almost equivalent results were obtained for Subotica, Belgrade, and Kopaonik, with the lowest potential found in Užice (see Table 4.20). The monthly electricity production for the considered PV systems for the case of Belgrade is shown in Fig. 4.22. Note that the solar energy production followed a similar trend in the other locations.

These results reveal the opportunities that exist for the production of PV electricity throughout Serbia, using both free-standing PV and the BIPV system. It should also be noted that when the solar energy generation exceeds the house's annual energy consumption, there are obvious opportunities for developing an energy plus home. Hence, these results may contribute to any guidelines for developing future energy plus housing concepts.

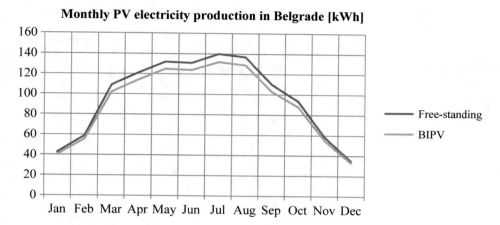

Fig. 4.22 Average monthly PV electricity production [kWh] for the case of Belgrade for the two considered systems, as determined by the PVGIS simulation tool (see Fig. 4.21)

4.7 Summary

The main topics of this chapter involved assessing the individual impact of seven building design parameters. The importance of each parameter was dependent upon the housing type being assessed, which operating regime (winter or summer) was considered, as well as the geographical region in the country, based on the five selected locations. The result was a series of patterns, ranked as recommended, acceptable, and unfavorable, for more efficient building design in Serbia. Such patterns would form the framework for guidelines for the early stages of a building's design, as well as for future architectural developments.

Energy efficiency is the starting point toward delivering nearly zero-energy and energy-plus buildings. In addition to high efficiency, a building should integrate the utilization of renewable energy sources. For example, PV energy production, as was examined in this chapter, was found to be a valuable option for efforts to deliver nearly zero-energy and energy-plus buildings in Serbia/Southeast Europe.

The next and final chapter summarizes the findings of the research carried out as part of this book. As these results represent simply another step toward the development of energy-efficient housing, future paths of research and the hope for the next generation of homes are also proposed.

References

BRE. (2011). Passivhaus primer: Designer's guide A guide for the design team and local authorities. Retrieved from http://www.passivhaus.org.uk/page.jsp?id=108, (accessed 15.02.2012).

IBO. (2013). Baubook Rechner für Bauteile, retrieved from https://www.baubook.at/, (accessed 10.02.2014).

PVGIS. (2014). Photovoltaic Geographical Information System, retrieved from http://re.jrc.ec.europa.eu/pvgis/ (accessed 01.12.2014).

Republic Hydro meteorological Service of Serbia. (2013). Retrieved from http://www.hidmet.gov.rs/, (accessed 15.01.2013).

Stamenić, Lj. (2010) Alternativna energija Srbije, Ekonomist Magazin br. 518 spec. dodatak, Politika a.d.

Conclusions: Energy-Efficient Buildings in Southeast Europe

5

5.1 The Design of More Energy-Efficient Homes

The intention of this book was to establish design patterns for more energy-efficient houses that would be suitable for the different regions of Serbia. Therefore, the work was divided into three main parts: first, a review of the traditional housing of Serbia (Chap. 2), then, an analysis of existing buildings and proposed effective refurbishment solutions (Chap. 3), and finally, to propose design patterns for future houses across different regions of the country (Chap. 4). For the design chapters, it was important to include not only the "architectural" issues that needed to be considered but also factors associated with the environment, economics, and social aspects. This chapter, the final, will review some of what has been presented in this book, as well as outline the sorts of design paths future homes in Serbia may follow to achieve greater energy efficiency.

5.2 Traditional Architecture

The first point to take away from a study of the traditional architecture of Serbia (or any region for that matter) is how the builders and residents took particular notice of their environment and surroundings. This was very apparent from the styles identified in the three Serbian regions considered, where different building techniques were developed to create a comfortable house from local available resources. However, it is not only the environment but the historical context that influenced the development of a region's architecture. For example, the way settlements developed was a product of a range of factors, such as the background of the settling population and the political context in terms of imposed regulations and fundamental security.

© Springer Fachmedien Wiesbaden GmbH, part of Springer Nature 2019
V. Jovanović, *Energy-efficient building design in Southeast Europe*,
https://doi.org/10.1007/978-3-658-24165-0_5

Nonetheless, the environment is the critical factor, especially when we again wish to consider modern concerns about energy efficiency. Such influences may manifest themselves in the most simple ways, for example, the steeper roofs employed by houses in the mountains to prevent the build-up of snow as opposed to flatter roofs in lower-lying areas. The available resources (while today perhaps not so critical) were a crucial factor, where large logs could be readily obtained from forested areas such as in the west, leading to the development of log cabins, or the rammed-earth housing of the north.

Therefore, it should be the architect's responsibility to address each location with a climate-appropriate design, according to the qualities of the location. Learning from the folk architecture therefore provides a good starting point toward environmentally friendly architecture while at the same time providing the required knowledge in the event that refurbishment of such dwellings is desired.

5.3   Refurbishment of Existing Buildings

It was readily seen early in this book that there is a huge potential for energy saving within the residential building sector of Serbia. The previous chapters outlined scenarios for refurbishing existing buildings (Chap. 3) and the better design of future ones (Chap. 4) within the context of improving energy efficiency, increased thermal comfort, and the consideration of renewable resources, specifically solar energy. The modeling was undertaken using input parameters that were relevant for the types of housing, both traditional and more modern, commonly found throughout Serbia, while accounting for geographical and climatic factors.

The first results (case study 1) dealt with historical building renovations (Sect. 3.5.1), where the potential for such refurbishments were found to be rather limited. Careful planning and design is required, in particular, so that no structural damage may arise (especially considering that such buildings are frequently protected by law). Hence, with any intended designs, not only energy efficiency but also structural-physical analyses need to be applied.

Next was an assessment (case study 2) of the refurbishment of multistory buildings (Sect. 3.5.2). What was soon realized was the immense potential for improved energy efficiency, by up to 90% in some cases, depending on the strategy followed. The assessed examples were a very compact building in a row, with only two façades, resulting in a relatively low amount of heat loss. The renovation of such buildings in fact offers a major opportunity for Serbian cities to reduce their energy usage and, as discussed latter, GHG emissions.

An important aspect of establishing a culture for both refurbishing and using improved designs for energy efficiency is the consideration of the socioeconomic context (Sect. 3.5.3). Hence, an analysis (case study 3) was carried out examining some of these within the context of Serbia and comparing them with other parts of the EU. Factors such as the cost-effectiveness of refurbishment in Serbia were similar to those in EU; in fact, owing to the lower required investments (generally due to lower labor costs), higher rates of energy saving, and shorter pay-off periods, these were found achievable within reasonable circumstances. However, several factors working against the establishment of such a culture were identified, which would play a major role is the decisions made by homeowners. The first was the disparity in local energy prices, which provided other avenues for reducing energy costs rather than refurbishment. For example, it was found to be cost-effective to change the heating source rather than retrofitting. Another issue was the actual ability of homeowners to afford the refurbishments, given the relatively low average income, which limited the feasibility of such investments.

The fourth case study dealt with how policy-driven refurbishments (Sect. 3.5.4) dictated the attractiveness of the proposed refurbishment choices. Pros and cons were identified when dealing with the environmental, economic, and social aspects of refurbishment. This included considering measures of energy efficiency (related to GHG emissions), financial benefit, and the resulting thermal comfort. Although the energy efficiency and thermal comfort aspects performed quite well compared to unmodified designs, the economic assessment exposed some barriers. For example, the passive house refurbishment scenario in Serbia was found to be an economically unattractive option.

The fifth study considered the effect of a series of design patterns, resulting in a kind of "toolbox" (Sect. 3.5.5), and what their effect was on houses dating from the 1970s and 1980 in Serbia, a period of rapid housing construction. The assessed refurbishment patterns determined the effectiveness of individual energy conservation measures, as well as their combined impact. The application of the appropriate patterns could provide energy savings of around 90%, the knowledge of which would help architects and planners apply these to future designs and to similar housing types.

As mentioned above, the choice of heating source potentially has a large impact of the costs of energy usage for a house. Case study 6 (Sect. 3.5.6) considered the replacement of a heating system in terms of energy efficiency and environmental impact, namely, their respective contributions to GHG emissions. It was found that while wood stoves (very common in Serbia) had a rather low energy conversion efficiency, the actual environmental impact in terms of GHG emissions was also relatively low. Hence, the need to install a more energy-efficient system while at the same time setting out to reduce one's GHG foot print is another factor requiring deeper consideration.

5.4   Designs for Future Houses

The next series of case studies were concerned with setting out proposals for future house designs that accommodated energy efficiency, thermal comfort, as well as possibilities for renewable energy, namely, solar energy (Chap. 4). One outcome was the finding that the prospects for so-called nearly zero-energy buildings to be built in Serbia are very good.

However, first a series of seven design parameters (compactness, standing, orientation, window glazing, insulation, construction type, prevention of overheating, Sect. 4.2) were selected for the assessment of housing designs. Then, the first study considering these factors, case study 7, was for a detached house (Sect. 4.3), where it was found that while some parameters, such as a house's compactness or its southern-facing orientation, had a relatively low influence on its energy efficiency, others such as its standing and level of insulation and ventilation equipment saw a greater influence in improving the house's performance. In particular, super insulation and HVAC equipment lead to a 90% decrease in heating demands, in comparison to what would be required if simply following the standard building code. However, the results did not always lead to situations where "simple" decisions could be made. For example, high levels of window glazing on the south-facing façade, while reducing heating demands under winter conditions, also increased the potential for rooms to overheat in summer, leading to the need for more air conditioning. Considering Belgrade as an example, only the options of 10 and 20% of glazing resulted in acceptable internal air temperatures in summer. With regard to suitable construction types, massive construction appears to perform better than the lightweight option in general, although this was not always the case. Moreover, passive cooling measures (e.g., shutters, night ventilation) appear to be essential.

The following case study (study 8) assessed the same seven parameters within the context of four types of residential buildings (three single-family houses and one multi-family six-story building). The results showed (Sect. 4.4) that each of the single-family houses required passive cooling measures, such as external shutters and increased night ventilation, while for the multistory residential building, the massive type of construction leads to a greater tendency for the overheating of rooms than it did for the single-family houses.

The next investigation, case study 9, involved design patterns that were suited to the different geographical regions of Serbia. Five locations, Subotica, Belgrade, Užice, Kopaonik, and Niš (Fig. 4.14), with their own local climatic conditions, landscapes, and traditional architecture, were selected. In each location, the level of insulation had the largest effect, hence being the key factor for designing energy-efficient houses. It was found that installing so-called super insulation and efficient HVAC equipment again allowed a 90% improvement relative to the basic model (as defined in the current building code). Small changes arose from altering the orientation of the south-facing façade, while the proportion of glazing showed a significant influence with regard to the potential for overheating in summer. The results of this assessment led to a series of patterns (Tables 4.14, 4.15, 4.16, 4.17 and 4.18), that represent recommended, acceptable, and inappropriate designs. These design patterns for five locations can therefore be used as a starting point for future designs for similar projects.

Finally, the prospects for the exploitation of renewable energy, especially solar energy (of which Serbia has abundance when compared to the rest of Europe), were considered (Case Study 10). Renewables (i.e., solar energy) were found to potentially be able to provide most energy demands of residential buildings, if considered from the beginning of the house's construction. Free-standing PV (photovoltaic) modules were found to be more effective than building-integrated photovoltaics (BIPV), with between 1040 and 1200 kWh per year potentially able to be produced for each of the five locations, indicating the possibilities for such decentralized renewable energy production in Serbia. It therefore provides another helpful direction that should be followed by Serbian architecture in the future.

While the developed patterns are not meant to replace the detailed studies required for a given project, nor are they intended to represent ideal architectural designs, their value lies in their flexibility, and that they are easily applicable. They may therefore be considered to represent a conceptual form for the design of energy efficiency and comfortable housing, providing a pre-step for the design of future housing in Serbia.

Every project is specific, and the best approach for an architect to follow is to analyze the current project in terms of desired or required building performance (as stated in Jovanović 2016). From this viewpoint, the design patterns are the starting point, as well as a sound source of information, especially if the architect is not performing simulations themselves, and is involved with similar building types to those considered in this book.

5.5   The Next Steps?

The aim of this book is to present an innovative approach for designing more energy-efficient buildings that are suitable for Southeastern Europe, with Serbia being the test case. Because of how the different geographical regions in Serbia extend beyond its borders, the specific issues analyzed in its case could be applied to other parts of the region.

However, the research that went into this work also raised additional questions. This involved revealing avenues through which future-related research may be applied and the current results improved upon. There were limitations in the research methodology followed, including evaluations based on single indicators, limited economic analysis, a lack of detailed design options, and a limited sample space. The undertaken evaluations were in turn based on rather specific indicators, namely, heating demands, internal air temperatures for thermal comfort, the designs' cost-effectiveness in terms of investment-return period, environmental impact as described by the CO_2 footprint, and the viability of renewable energy by considering PV electricity potential. This approach was intentional so as to deal with the most critical issues surrounding the building of more energy-efficient and comfortable houses in the future. As one would imagine, to comprehensively assess all of these and the accompanying more detailed issues would be outside the scope of this book. Similarly, while some economic assessments were made for the refurbishment studies, they were neglected for the future designs investigations. The reason for this was simply because a full economic analysis would require the collection of massive amounts of information from construction companies, an impractical undertaking for a book of this nature.

While this book did not go into very technical details about each of the design parameters assessed, more thorough comparisons considering this would allow an expanded view of how more energy-efficient housing may be designed. For example, taking the case of glazing, the optimal level of glazing per façade could be well defined by considering the balance between the needs and requirements for winter warming and summer cooling, an issue that arose during several of the case studies. While the software tools employed could not assist in this (the specific simulation software used did not allow cooling demands to be outputted), solutions were nonetheless proposed that considered cooling demands, for example, options such as night ventilation.

Another point concerns expanding the number of cases examined. Considering the dependency on geographical location, while examples of typical structures were examined for five locations around Serbia, greater generality could be achieved by expanding the number of building types studied. Nonetheless, the selected examples of buildings for the simulations were carefully chosen so that the findings of the case studies could be easily applied to other similar building types as well as being relevant to a relatively large target group. In addition, the analyses were carried out considering both winter and summer conditions, which is essential for Serbia considering possible conflicting needs, as the case of window glazing showed. Furthermore, in order to develop replicable conceptual models, the research approach was adapted and expanded from previous studies for Serbia (Jovanović Popović and Ignjatović 2012) and other regions (Bointner et al. 2012; Konstantinou 2014).

Given that this book dealt exclusively with residential buildings, further research should address the design of large-scale architecture. This would include, and require, other indicators of sustainable design to be incorporated. Such studies would in turn lead to the ability to update and expand upon guidelines for energy-plus buildings at the regional scale in Europe.

While deficiencies recognized during this work have been mentioned, it is worth stating that some gaps from previous studies were dealt with. For example, some of the neglected issues in the past research include economic feasibility, summer time overheating potential, and developing replicable conceptual models (Pucar 2006; Bojić et al. 2012; Stevanović et al. 2009). In particular, economic assessments of specific scenarios of renovations were examined, given that this is a serious limiting factor in encouraging more widespread retrofitting in Serbia. Other issues studied included socioeconomic factors within the context of the local conditions, especially those related to the actual affordability of any required investments.

Finally, the investigated state, Serbia, is a centrally positioned Southeastern European country. Therefore, as mentioned throughout this book, the varied geography within the country spreads out throughout the region, allowing the exporting of the developed design patterns to neighboring countries. Meanwhile, Serbia is making efforts to join the EU and is expected to be a member by 2025. As a result, due to the country's progress in harmonizing its energy policy with that of EU legislative, energy efficiency is becoming a priority for the Serbian building industry. Furthermore, this issue itself opens avenues of study, and while doing so, permitting the dissemination of ideas about more efficient building designs, such as those outlined in this book.

5.6   Final Word: The Home of the Future

This book has provided an example of a framework and workflow for integrating energy-efficiency principles into the conventional building design process. This was done by early-design analysis of climate-appropriate building envelopes and their thermal properties, within the context of Southeast Europe, in particular the case of Serbia.

The research undertaken that formed the basis for this book resulted in many recommendations for future building design in Serbia. Several steps are recognized for these recommendations to be most effectively employed. First, at the most basic level, each considered design must include a detailed assessment of the environmental, economic, and social contexts of the building to be constructed. Next, accompanying the whole process is the need for the relevant professions to make recommendations and be involved in the updating and refinement of local and national energy policy. For example, this could see the development of a strategy for implementing a platform that oversees regional energy-efficient designs, bringing together all relevant stakeholders (architects, engineers, environmental scientists, economists, social scientists, land-use planners, and of course, the wider public). This would include examining further the socioeconomic issues surrounding the acceptance and implementation of retrofitting for more energy-efficient homes. It would also need to make the consideration of integrated energy efficiency more widespread (in addition to what is defined by current building codes) and a "standard" factor when designing a home, from both the architect's and home owner's perspectives. What cannot be neglected is that the traditional homes that still exist throughout the country were the result of different building techniques that evolved to create comfortable houses suitable for the local environment and available resources. Therefore, it is the architect's responsibility to take into account the geographical specifics when designing a structure. This leads to the need to learn from folk architecture, which provides a sound start toward the development and construction of environmentally friendly and appropriate architecture. However, coupled with this is the importance of recognizing the special case of historical architecture when implementing refurbishment schemes to meet modern expectations of energy efficiency.

It is hoped that the ideas and suggestions proposed in this book will contribute to remedying the practical problems faced by many professionals in the construction industry. For example, the described design patterns are considered to be of benefit to designers, due to the high impact of the evaluated parameters on a building's overall energy performance. Hence, the presented design patterns are proposed as open concepts and are hoped to serve as a starting point for energy-efficient housing design in different regions in Serbia. This would allow architects to integrate carefully considered combinations of measures into the very early design phase and, therefore, strongly influence the overall energy balance of the building. Although the future is unknown, in terms of cost-effective renovations, appropriate designs and, not least, the exploitation of renewable energy resources such as solar energy, the Southeast European region, and Serbia in particular, should have every reason to have confidence in the opportunities available for the development of sustainable energy-positive architecture.

References

Jovanović, V. (2016). Improving integrated building design with early design simulations. Proceedings Advanced Building Skins Conference, Bern – Switzerland.

Jovanović Popović, M., & Ignjatović, D. (2012). Atlas of Family Housing in Serbia. Faculty of Architecture, University of Belgrade and GIZ–German Association for International Cooperation, Belgrade.

Bointner R. et al. (2012). Gebäude maximaler Energieeffizienz mit integrierter erneuerbarer Energieerschließung, Berichte aus Energie-und Umweltforschung 56a/2012. Bundesministeriums für Verkehr, Innovation und Technologie, Wien, 2012.

Konstantinou, T. (2014). Facade Refurbishment Toolbox: Supporting the Design of Residential Energy Upgrades. Delft University of Technology, Faculty of Architecture and The Build Environment, Architectural Engineering + Technology department.

Pucar, M. (2006). Bioklimatska arhitektura, zastakljeni prostor i pasivni solarni sistemi, Beograd.

Stevanović S., Pucar M., Kosorić V. (2009). Potential solar energy use in a residential district in Niš. Spatium.

Bojić, M., et al. (2012). Decreasing energy consumption in thermally non-insulated old house via refurbishment. Energy and Buildings.